20th International Conference on Electron Dynamics in Semiconductors, Optoelectronics and Nanostructures 2017 (EDISON 20)

AA006285

Journal of Physics: Conference Series Volume 906

Buffalo, New York, USA
17 - 21 July 2017

Editors:

Erik Einarsson

ISBN: 978-1-5108-5037-8
ISSN: 1742-6588

Printed from e-media with permission by:

Curran Associates, Inc.
57 Morehouse Lane
Red Hook, NY 12571

CURRAN ASSOCIATES INC. proceedings .com

Some format issues inherent in the e-media version may also appear in this print version.

Printed by Curran Associates, Inc. (2017)

For permission requests, please contact the Institute of Physics
at the address below.

Institute of Physics
Dirac House, Temple Back
Bristol BS1 6BE UK

Phone: 44 1 17 929 7481
Fax: 44 1 17 920 0979

techtracking@iop.org

Additional copies of this publication are available from:

Curran Associates, Inc.
57 Morehouse Lane
Red Hook, NY 12571 USA
Phone: 845-758-0400
Fax: 845-758-2633
Email: curran@proceedings.com
Web: www.proceedings.com

Table of contents

Volume 906

20th International Conference on Electron Dynamics in Semiconductors, Optoelectronics and Nanostructures (EDISON 20)

17–21 July 2017, Buffalo, New York, USA

Accepted papers received: 6 October 2017
Published online: 22 October 2017

Preface

Papers

IOP Conf. Series: Journal of Physics: Conf. Series **906** (2017) 011001 doi:10.1088/1742-6596/906/1/011001

The 20th *International Conference on Electron Dynamics in Semiconductors, Optoelectronics and Nanostructures* (EDISON 20) was held at the Hyatt Regency Hotel and Conference Center in downtown Buffalo, New York, USA from July 17–21, 2017. The meeting serves as the premier forum for researchers working on the study of nonequilibrium phenomena in semiconductors, from all parts of the globe, and was organized locally by faculty at the University at Buffalo and the University of Rochester.

EDISON is an international conference series with biennial meetings that cover recent progress in the field of electron dynamics in solid-state materials and devices. Initially established with the title *Hot Carriers in Semiconductors (HCIS),* the first meeting took place in Modena, Italy in 1973. The name was changed to *International Conference on Nonequilibrium Carrier Dynamics in Semiconductors* at the 1997 Berlin meeting, and again at the 2009 meeting in Montpellier to the current form. Since then, EDISON conferences have been held in Santa Barbara, USA (2011); Matsue, Japan (2013); Salamanca, Spain (2015); and Buffalo, USA (2017).

The EDISON 20 program included 12 invited talks, 50 oral contributions, 116 accepted abstracts, and one special seminar. In total, 152 participants from 22 countries gathered to share and exchange ideas and recent results. Approximately 45% of the participants were from North America, 30% from Europe, and 25% from the Asia Pacific region. Fully 30% of the participants were students, and there was a very good balance between experimental and theoretical approaches. Topics with the most contributions were: Electronic and optical properties of graphene and other 2D materials; Nonequilibrium carrier transport; and Terahertz phenomena in semiconductor materials and devices.

I thank all the members of the Program Committee and the International Advisory Committee for their valuable input, as well as Professor Jonathan Bird and Alex Desha for their many hours devoted to organizing a successful EDISON conference. Also critical to the success of the conference is financial support, so I thank the School of Engineering and Applied Sciences, the Vice President for Research and Economic Development, the Department of Physics, and the Department of Electrical Engineering at the University at Buffalo, as well as the U.S. Department of Energy, Kurt J. Lesker Company, and the Hajim School of Engineering & Applied Sciences at the University of Rochester for their sponsorship of EDISON 20. Finally, I thank all the scientists who contributed to the preparation and peer-review of these proceedings.

Erik Einarsson, Editor
EDISON 20 Publications Chair
University at Buffalo

IOP Conf. Series: Journal of Physics: Conf. Series **906** (2017) 011002 doi:10.1088/1742-6596/906/1/011002

Peer review statement

All papers published in this volume of *Journal of Physics: Conference Series* have been peer reviewed through processes administered by the proceedings Editors. Reviews were conducted by expert referees to the professional and scientific standards expected of a proceedings journal published by IOP Publishing.

IOP Conf. Series: Journal of Physics: Conf. Series **906** (2017) 012001 doi:10.1088/1742-6596/906/1/012001

Low-temperature performance of semiconducting asymmetric nanochannel diodes

Y Akbas[1], G R Savich[1], A Jukna,[1,2] T Plecenik[3], P Ďurina[3], A Plecenik[3], G W Wicks[1] and Roman Sobolewski[1]

[1]University of Rochester, Rochester, NY 14627-0231, USA
[2]Vilnius Gediminas Technical University, Vilnius LT-10223, Lithuania
[3]Comenius University, SK-84248 Bratislava, Slovakia

E-mail: roman.sobolewski@rochester.edu

Abstract. We present our studies on fabrication and electrical and optical characterization of semiconducting asymmetric nanochannel diodes (ANCDs), focusing mainly on the temperature dependence of their current–voltage (*I–V*) characteristics in the range from room temperature to 77 K. These measurements enable us to elucidate the electron transport mechanism in a nanochannel. Our test devices were fabricated in a GaAs/AlGaAs heterostructure with a two-dimensional electron gas layer and were patterned using electron-beam lithography. The 250-nm-wide, 70-nm-deep trenches that define the nanochannel were ion-beam etched using the photoresist as a mask, so the resulting nanostructure consisted of approximately ten ANCDs connected in parallel with 2-μm-long, 230-nm-wide nanochannels. The ANCD *I–V* curves collected in the dark exhibited nonlinear, diode-type behavior at all tested temperatures. Their forward-biased regions were fitted to the classical diode equation with a thermionic barrier, with the ideality factor n and the saturation current as fitting parameters. We have obtained very good fits, but with n as large as ~50, suggesting that there must be a substantial voltage drop likely at the contact pads. The thermionic energy barrier was determined to be 56 meV at high temperatures. We have also observed that under optical illumination our ANCDs at low temperatures exhibited, at low illumination powers, a very strong photoresponse enhancement that exceeded that at room temperature. At 78 K, the responsivity was of the order of 10^4 A/W at the nW-level light excitation.

1. Introduction

Intrinsic nanostructures exhibit unique physical properties that are a direct consequence of their nanoscale parameters and the quantum mode of operation, in contrast to scaled-down versions of well-known conventional devices. One of the best-known examples is a nonlinear, asymmetric nanochannel diode (ANCD), also called a self-switching diode (SSD), invented in 2004 by Song *et al.* [1]. Unlike conventional (e.g., *p-n*) diodes, the ANCD performance is not based on the junction barrier; instead, it produces diode-like characteristics through nonlinear carrier transport in a depletion-controlled nanochannel. Since depletion control is purely geometrical and the device itself has a planar geometry, there is room for flexibility in terms of material selection and device design. In terms of the applications, when combined with coupling antennas, ANCDs act as reliable and sensitive THz detectors [2], as well as (based on Monte Carlo simulations) THz generators [3]. They have also been

IOP Conf. Series: Journal of Physics: Conf. Series **906** (2017) 012001 doi:10.1088/1742-6596/906/1/012001

proposed as ultrafast logic gate operators [4] and most recently have been demonstrated as ultrasensitive optical detectors [5].

To maximize carrier transport, most ANCDs are fabricated on wafers with a two-dimensional electron gas (2DEG) layer. Therefore, the ballistic transport enables their ultrafast operation or, equivalently, THz-frequency bandwidth. Our research presented here focuses on exploiting low-temperature performance ANCDs with strongly nonlinear current–voltage (*I–V*) characteristics and demonstrates that they exhibit a very large gain at ultralow optical excitations [5]. The latter makes them promising for single-photon detection applications.

2. Fabrication and Experimental Setup

A schematic cross section of an actual wafer used to fabricate our devices is shown in Figure 1a. It is a GaAs/AlGaAs quantum-well heterostructure grown in-house by molecular-beam epitaxy on a semi-insulating GaAs substrate that exhibits a 2DEG layer at the GaAs/AlGaAs interface, ~50 nm below the surface. The heterostructure cross section in Figure 1a is overlaid with an atomic force microscopy (AFM) profile from an actual fabricated device, showing that the ~65-nm trench etching is deep enough to isolate the 2DEG layer. A sketch of a fabricated device chip is shown in Figure 1b. The entire structure consists of contact pads (blue areas) with 20-nm-thick Pt metallization and a mesa structure (gray region) etched out to isolate the ANCDs from their surroundings. Finally, the actual nanostructures (black U shapes) are fabricated at the mesa center. Using electron-beam lithography patterning and the resist as a mask, nanometer-width trenches are etched down below the 2DEG level (see Figure 1a) via ion-beam etching. Figure 1c presents an AFM top view image of completed ANCDs. For practical purposes, the ends of trenches bend away from each other. Typically, about ten devices connected in parallel were fabricated, each having a <2-μm-long, ~230-nm-wide nanochannel. The trenches were ~65 nm deep (see Figure 1a).

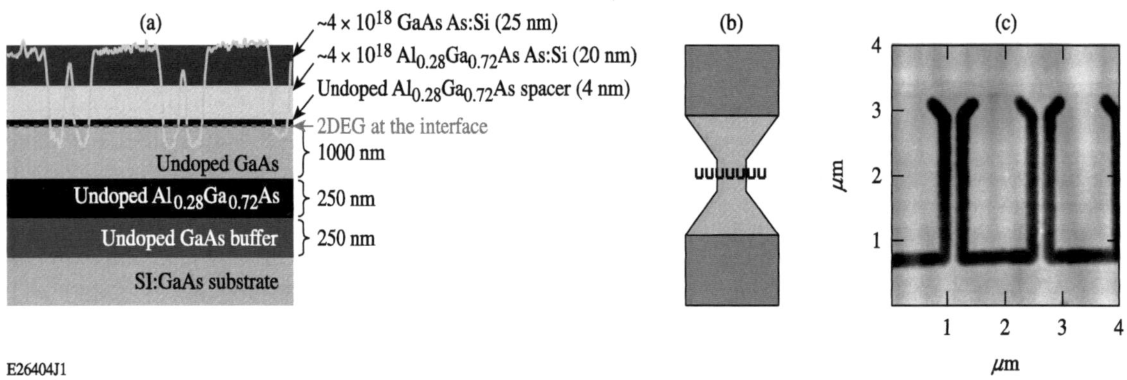

Figure 1. (a) Schematics of a GaAs-based heterostructure with a 2DEG layer ~50 nm below the surface. The heterostructure cross section has been overlaid with an AFM profile of the actual etched ANCD nanostructure. (b) Sketch of a test structure with ANCDs in parallel at the centre of the mesa (black U-shaped structures). (c) An AFM image of a section of the completed ANCDs.

The ANCD *I–V* characteristics have been collected by biasing the device with a constant-voltage source and measuring the transport current under both the dark and light illumination conditions. The device under test was placed inside a liquid nitrogen cryostat with glass windows for optical access, enabling us to perform measurements in the temperature range between 78 K and 300 K. The cryostat also acted as a shield to block stray light. For optical excitation experiments, we used a train of 800-nm-wavelength, 100-fs-wide pulses of light generated by a commercial Ti:sapphire laser with a 76-MHz repetition rate. Using a bank of neutral-density attenuators, different excitation levels from 1 mW to 1 nW could be achieved. A spot size of optical excitation was selected to be 20 μm to ensure uniform optical power delivery to our ANCDs connected in parallel.

3. Results and Discussion

The smaller of the two insets in Figure 2 presents a set of the ANCD *I–V* curves collected at different temperatures *T* between room and liquid nitrogen temperatures. We note that our device exhibits highly nonlinear characteristics and the nonlinearity, as well as the asymmetry of the *I–V* curve, increases with a decrease in temperature. In addition, above ~2.5 V of bias, the traces level off as electrons in the nanochannel reach saturation with a velocity of the order of 10^7 cm/s, corresponding to the ballistic transport. Interestingly, the saturation current level does not simply scale with temperature, which will require further research. At the negative bias, below ~ –2.5 V, the leakage current starts to increase, likely because of the ballistic transport along, this time, a depleted nanochannel.

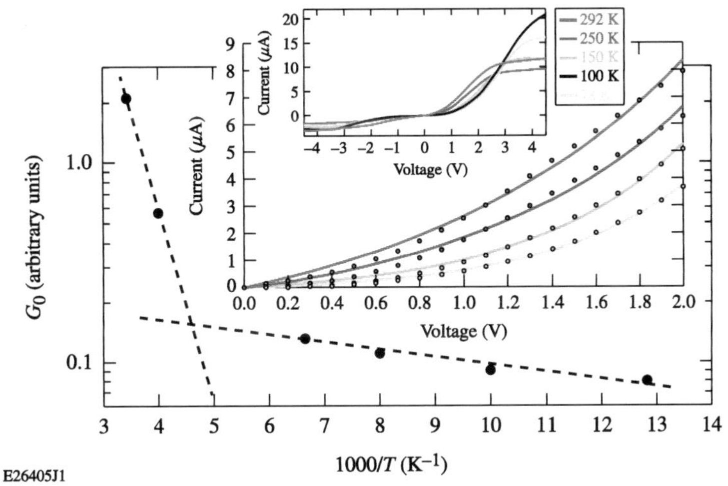

Figure 2. Differential conductivity G_0 plotted versus reciprocal temperature (large black dots) with dashed black lines indicating two regimes of the thermionic electron transport. The smaller inset presents *I–V* characteristics of our ANCD measured at different temperatures (see legend). The larger inset shows forward-biased parts of the same characteristics as the ones in the smaller inset (dots) overlaid with numerical fits (solid lines) based on Eq. (1).

The larger inset in Figure 2 presents forward-biased parts (0- to 2-V regions; dots) of the same *I–V* characteristics as shown in the smaller inset, but now overlaid with fits (solid lines) based on the diode equation analog to the Schottky junction equation, i.e., the one that includes a thermionic barrier:

$$I = I_0 e^{-\phi_{\text{barrier}}/kT}\left(e^{qV/nkT} - 1\right), \quad (1)$$

where I_0 is a constant corresponding to the saturation current, ϕ_{barrier} is the thermionic barrier energy, and *n* is the ideality factor. Clearly, the fits are very good at all tested temperatures ($R^2 > 0.99$); however, we had to use *n* as large as ~50. The latter is physically unrealistic and must indicate that there is a very large voltage drop at the contact pads, or, possibly, it might be related to an extra voltage drop across the trenches.

Finally the zero-bias differential conductance $G_0 = \mathrm{d}I/\mathrm{d}V$ at $V = 0$ (main panel in Figure 2) is extracted from each *I–V* curve and plotted as a function of reciprocal temperature in a semi-logarithmic scale. The observed behavior indicates the existence of a thermionic energy barrier with ϕ_{barrier} = 56 meV at high temperatures that becomes strongly suppressed (ϕ_{barrier} = 10 meV) below ~200 K. Therefore, we believe that at temperatures close to room temperature and near-zero bias, our depleted nanochannel represents a thermionic barrier that can be spontaneously/thermally negotiated by carriers as the bias voltage increases, in a way similar to the case of Schottky junctions. At low temperatures, the barrier is much lower and transport becomes ballistic.

Following our earlier research [5], we also performed optical characterization of our ANCDs. We measured the photocurrent's dependence on the incident optical power and then calculated the responsivity. Analogous to our earlier observations, the devices presented here also exhibit a dramatic increase in responsivity as the incident optical power is decreased. The same trend is observed at both room and low temperatures; however, at low temperatures the responsivity increase becomes a steeper function of the optical power, as is shown in Figure 3. We expect that at the single-photon power level (SP point in Figure 3), the responsivity should reach 25000 A/W at 78 K, which should be well within the reach of our next-generation ANCDs. The single-photon level has been calculated to be ~19 pW, corresponding to an average of one photon per 100-fs pulse with an 800-nm wavelength and a 76-MHz repetition rate generated by our laser. Equivalent detection efficiency indicates that for a single absorbed photon, ~1500 electrons should pass through the nanochannel.

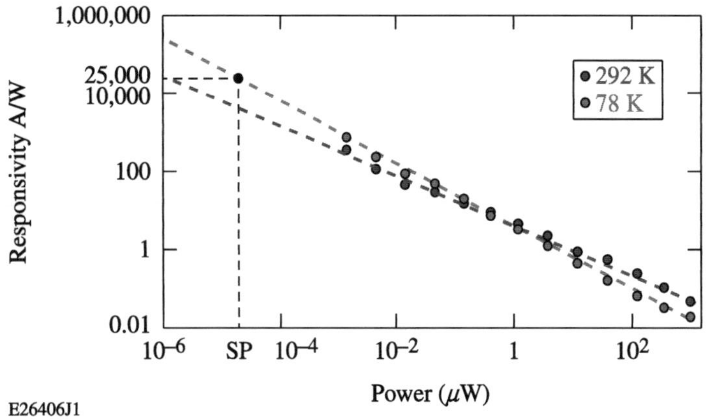

Figure 3. Responsivity versus incident optical power for an ANCD tested at room and liquid nitrogen temperatures (dots). The dashed lines are the linear fits to experimental values. The black dot represents the calculated single-photon (SP) illumination level.

4. Conclusions

We have performed systematic studies of the low-temperature performance of ANCDs fabricated in the GaAs/AlGaAs heterostructure with a 2DEG layer. Our ANCDs were characterized by nonlinear *I–V* characteristics that could be very well fitted using the standard diode equation with a thermionic barrier term. The zero-voltage differential conductivity extracted from the *I–V* curves demonstrated that at high temperatures ϕ_{barrier} = 56 meV and decreased substantially (to the 10-meV value) at temperatures below ~200 K. Our optical excitation experiments confirmed the earlier-observed existence of dramatic responsivity enhancement at very low optical powers incident upon the device; this effect was somewhat enhanced at low temperatures. The measured responsivity reached ~10^4 A/W at nW-level optical excitation at 78 K. With strongly suppressed dark currents at low temperatures, cryogenic ANCDs look to be promising candidates for single-photon–level detectors.

Acknowledgements
The authors thank M. Mosko for valuable discussions. This work was supported in part by the PumpPrimerII Program at the University of Rochester. AJ thanks the Baltic-American Freedom Foundation for their financial support.

References

[1] Song A M, Maximov I, Missous M and Seifert W 2004 *Physica E* **21** 1116
[2] Balocco C, Kasjoo S R, Lu X F, Zhang L Q, Alimi Y, Winnerl S and Song A M 2011 *Appl. Phys. Lett.* **98** 223501
[3] Mateos J, Vasallo B G, Pardo D and González T 2005 *Appl. Phys. Lett.* **86** 212103.
[4] Song A M, Missous M, Omling P, Maximov I, Seifert W and Samuelson L 2005 *Appl. Phys. Lett.* **86** 042106
[5] Akbas Y, Stern A, Zhang L Q, Alimi Y, Song A M, Iñiguez-de-la-Torre I, Mateos J, González T, Wicks G W and Sobolewski R 2015 *J. Phys.: Conf. Ser.* **647** 012013

IOP Conf. Series: Journal of Physics: Conf. Series **906** (2017) 012002 doi:10.1088/1742-6596/906/1/012002

Terahertz time-domain spectroscopy characterization of carbon nanostructures embedded in polymer

G Chen,[1] R Shrestha,[1] A Amori,[1] Z Staniszewski,[2] A Jukna,[1,3] A Korliov,[4] C Richter,[5] M El Fray,[2] T Krauss,[1] and Roman Sobolewski[1]

[1]University of Rochester, Rochester, NY 14627-0231, USA
[2]West Pomeranian University of Technology, Szczecin, PL-70322, Poland
[3]Vilnius Gediminas Technical University, Vilnius LT-10223, Lithuania
[4]Centre for Physical Sciences and Technology, Vilnius LT-10257, Lithuania
[5]University of Iceland, Reykjavik, Iceland

E-mail: roman.sobolewski@rochester.edu

Abstract. This work presents results of terahertz time-domain spectroscopy (THz-TDS) investigations of two types of polymer-based nanocomposites, consisting of bulk samples of graphene nanoflakes embedded in a polymer matrix and thin films where single-walled carbon nanotubes (SWCNTs) with a chiral index (7,5) were wrapped in single strains of a polymer. Our THz-TDS setup is a room-temperature system with a frequency range of 0.1 to 3.5 THz, based on photoconductive switches as both the emitter and detector, excited by 100-fs-wide optical pulses. We have studied THz spectra of both types of samples and compared them to the ones obtained for the pure polymer reference specimens and fitted the experimental complex conductivity data to the Drude–Smith model. The Drude–Smith model fits our data well, demonstrating that graphene nanoflakes and SWCNTs, exhibit highly localized intra grain/tube electron backscattering with a femtosecond relaxation time.

1. Introduction

The terahertz (THz) region of the electromagnetic spectrum has been receiving increasing interest in many areas of science, such as physics, materials science, chemistry, and biology, and has found many applications as a nondestructive and noncontact technique applicable in research laboratories, industrial production lines, security systems, and even ultrafast communications [1]. Specific spectral signatures [2] of a number of polar and crystalline materials [3], organic molecules, and solid and liquid composites [4] that are present in the far-infrared radiation region can be observed by means of transmission THz time-domain spectroscopy (THz-TDS). Since this technique simultaneously provides information on both the amplitude and phase of the THz transient, it can reveal extremely small changes in the complex index of refraction, or equivalently the conductivity, of various, e.g., composite materials [4]. Finally, because of the extremely high sensitivity of THz radiation to water, the THz-TDS method can detect even very minute amounts of water in, e.g., biological specimens [5].

The THz-TDS technique is also uniquely suitable for broadband, THz-range frequency characterization of semiconducting and dielectric materials and recently has been widely implemented in studies of carbon nanocomposites containing graphene and carbon nanotubes [6–9]. While is it well known that the presence of graphene nanoflakes improves the mechanical strength [10] and/or

IOP Conf. Series: Journal of Physics: Conf. Series **906** (2017) 012002 doi:10.1088/1742-6596/906/1/012002

biological sensing properties [11] of a polymer, it is also known that they affect dielectric properties of polymer matrices, resulting in a modified, THz-frequency dielectric function [7] or, equivalently, complex conductivity. A single-walled carbon nanotube (SWCNT) is a two-dimensional (2-D) graphene sheet rolled up into a hollow cylinder. The diameter and helicity of a defect-free SWCNT are characterized by the chiral vector connecting equivalent crystalline sites on a 2-D graphene sheet [6] and represented by chiral indexes, i.e., a pair of integers (n, m) that define how a graphene sheet can be rolled up into an individual SWCNT. Knowing the indices of a given SWCNT, one can determine its electronic band structure and electric conductivity [12], i.e., whether a given tube is a metal or a semiconductor. Therefore, SWCNTs in their different crystallographic forms have been considered attractive candidates for a number of diverse electronic applications such as molecular electronics, photovoltaic devices, field-effect transistors, etc.

2. Sample preparation and THz-TDS measurement setup

We have tested two types of polymer-based carbon nanocomposites: bulk samples of the graphene nanoflakes embedded in a poly(ethylene terephthalate)-based copolymer matrix (graphene+PET) [13] and thin films of SWCNTs wrapped with poly(9,9-di-n-octylfluorenyl-2,7-diyl) (PFO) strains, deposited by spin coating onto 1-mm-thick fused-silica substrates (SWCNTs+PFO). The bulk samples were prepared by intermixing a commercial, exfoliated graphene during a melt polymerization process and a subsequent hot-press technique. The resulting samples were polymer foils with a thickness ranging from 0.15 to 1.0 mm with 1% graphene content by weight. In the case of thin-film samples, they were ~400-nm-thick films and contained more than 90% of SWCNTs of chirality (7,5) with a tube diameter of 0.83 nm [14]. Consequently, they were expected to exhibit semiconducting electronic properties. Pure bulk PET polymer and thin PFO film samples of corresponding thicknesses were also fabricated and used as references.

Our THz-TDS setup is based on low-temperature–grown GaAs photoconductive switches implemented as both the THz transient emitter and detector. As an optical excitation, we use 800-nm-wavelength, 100-fs-wide optical pulses generated at a 76-MHz repetition rate by a commercial mode-locked Ti:sapphire laser. The emitter and detector were equipped with Si lenses, separated by 24 cm, and placed inside a dry nitrogen gas enclosure. The THz pulse spectrum routinely obtained in an empty setup (no sample between the emitter and detector) extended to 3.5 THz (see figure 1a; black curve).

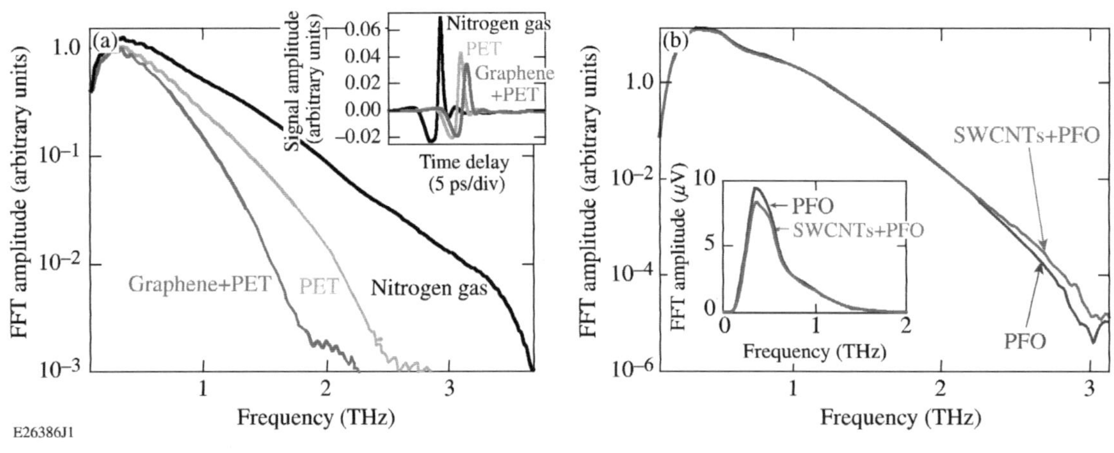

Figure 1. Semi-log plots of the transmitted THz signal spectra through (a) a 0.9-mm-thick foil of graphene+PET material (red line) and a reference (light blue line), and (b) a 0.4-mm-thick film of (7,5) SWCNT+PFO (red line) and a pure PFO reference spectrum (blue line). The inset in (a) shows experimental time-domain transients used to compute the spectra presented in the main panel. The inset in (b) shows the same data as the main panel, but the FFT amplitude is in a linear scale. In (a) we also present the spectrum of an empty setup (nitrogen gas, black line).

3. Measurement results and discussion

Figures 1a and 1b present spectra for both the graphene+PET and SWCNTs+PFO samples, as well as the spectra of corresponding, pure polymer references, obtained by means of a fast Fourier transform (FFT) from corresponding experimental time-domain transients (see examples in the inset in figure 1a). For both samples, the spectra are rather similar, i.e., the FFT amplitude of the THz signal decreases with increasing frequency, reaching the system noise floor level. However, we observe that for the selected 0.9-mm-thick graphene+PET sample, over the entire frequency range, the absorption is significantly larger than that for the pure PET sample. Conversely, for the SWCNTs+PFO film, its absorption spectrum almost overlaps with that of the pure PFO. There are, in this case, two noticeable exceptions: namely, an extra absorption at low frequencies with the peak around 0.4 THz (see inset in figure 1b) and a measurable absorption decrease for >2 THz. The latter effect could be related to an increased THz transmittance of SWCNTs at high frequencies [15].

Subsequently, we extracted the THz-frequency–range index of refraction $n(\omega)$ for both sample types using the relationship [16]

$$n(\omega) = 1 + \frac{c\left(\phi_{\text{sample}} - \phi_{\text{ref}}\right)}{\omega d}, \tag{1}$$

where c and d are the speed of light and sample thickness, respectively, ω is the angular frequency, and ϕ_{ref} and ϕ_{sample} are the phases of the reference and sample signals, respectively. In the case of the graphene+PET sample, the n spectrum is flat in the entire tested frequency range, but the presence of graphene flakes in the polymer increases its value from 1.75 to 1.85. The presence of nanotubes in a SWCNTs+PFO sample was more visible since the value of n roughly doubled, reaching ~20 at high frequencies.

Next, complex conductivity spectra for both sample types were calculated and are presented in figure 2 together with numerical fits (solid black lines) based on the Drude–Smith model [17], i.e.; the Drude conductivity with the Smith correction (second expression in parenthesis):

$$\sigma(\omega) = \frac{ne^2\tau/m}{1+\omega^2\tau^2}\left[1 + \frac{c\left(1-\omega^2\tau^2\right)}{1+\omega^2\tau^2}\right], \tag{2}$$

where n, m, and τ are the effective density, mass, and relaxation time of electrons, respectively, and c is the fitting parameter expressing a fraction of the electrons' original velocity that is retained after electron-scattering events.

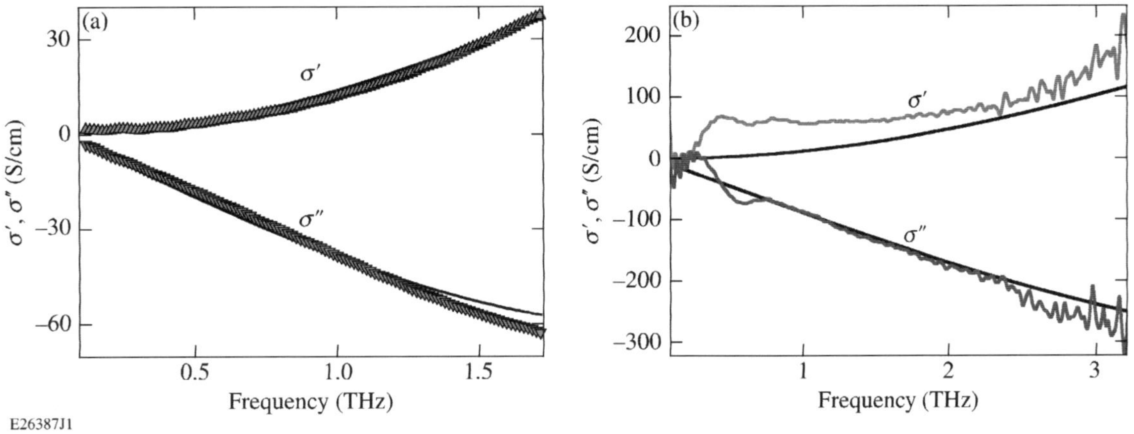

Figure 2. Real σ' and imaginary σ'' parts of the complex conductivity for (a) graphene+PET and (b) SWCNTs+PFO thin films. Black lines represent the Drude–Smith model fits [see Eq. (2)].

We note that for both carbon nanostructures, σ'' are negative and best fits shown in figures 2a and 2b correspond to $c \approx -1$, which indicates virtually complete elastic backscattering of electrons with relaxation times τ equal to 21 fs and 11 fs, for graphene and SWCNTs, respectively. A likely explanation is that carriers are fully localized within a given nanostructure and can freely move only inside the nano-carbon material.

4. Conclusions

Our results demonstrate that THz-frequency dielectric properties of nanocarbon-embedded polymer nanocomposites are sensitive to their nanocarbon forms, i.e., graphene nanoflakes or SWCNTs. In both tested samples, we have observed an enhanced index of refraction, as compared to pure polymer references; the effect was particularly large in the SWCNT+PFO sample. Complex conductivity spectra of both sample types exhibited the negative imaginary part and the $\sigma(\omega)$ experimental results could be very well fitted with a simple Drude–Smith model, indicating a total confinement of electrons inside localized carbon nanostructures with the femtosecond-range backscattering of carriers. A direct comparison of transmitted THz spectra of the test and reference samples enabled us to analyze frequency-dependent complex conductivity without resorting to Kramers–Kronig relations.

Acknowledgements

This work was supported in part by the PumpPrimerII Program at the University of Rochester. AJ thanks the Baltic-American Freedom Foundation for their financial support.

References

[1] Siegel P H 2004 *IEEE Trans. Microw. Theory Techn.* **52** 2438
[2] Kumar S, Kamaruju N, Karthikeyan B, Tondusson M, Freysz E, Sood A K 2010 *J. Phys. Chem. C* **114** 12446
[3] Wietzke S, Jansen C, Reuter M, Jung T, Kraft D, Chatterjee S, Fischer B M, Koch M 2011 *J. Mol. Struct.* **1006** 41
[4] Naftaly M and Miles R E 2007 *Proc. IEEE* **25** 1658
[5] Boyd J E, Briskman A, Colvin V L, Mittleman D M 2001 *Phys. Rev. Lett.* **87** 147401
[6] Peng X and Wong S S 2009 *Adv. Matter.* **21** 625
[7] Kampfrath T, von Volkmann K, Aguirre C M, Desjardins P, Martel R, Krenz M, Frischkorn C, Wolf M and Perfetti L 2008 *Phys. Rev. Lett.* **101** 267403
[8] Xu X, Chuang K, Nicholas R J, Johnston M B and Herz L M 2009 *J. Phys. Chem. C* **113** 18106
[9] Luo L, Chatzakis I, Patz A and Wang J 2015 OSA Technical Digest (online) (Washington, DC: *Optical Society of America*), Paper FM1B.1
[10] See, e.g., Shokrieh M M, Hosseinkhani M R, Naimi-Jamal M R, Tourani H 2013 *Polym. Test.* **32** 45
[11] Sobolewski P, Piwowarczyk M, El Fray M 2016 *Macromol. Biosci.* **16** 944 and references therein
[12] Bandaru P R 2007 *J. Nanosci. Nanotechnol.* **7** 1
[13] Staniszewski Z and El Fray M 2016 *Polimery* **61** 482
[14] Qin L-C 2007 *Phys. Chem. Chem. Phys.* **9** 31
[15] Wachi Y, Kawamoto T, and Sugiyama Y 2012 US Patent No. 2012/0252377 A1
[16] Fischer B M, Helm H and Jepsen P U 2007 *Proc. IEEE* **95** 1592
[17] Smith N V 2001 *Phys. Rev. B* **64** 155106

Experimental and theoretical studies of Sub-THz detection using strained-Si FETs

J.A. Delgado Notario[1], E. Javadi[1,2], V. Clericò[1], K. Fobelets[3], T. Otsuji[4], E. Diez[1], J.E. Velázquez-Pérez[1], and Y.M. Meziani[1]

[1]NanoLab, Salamanca University, 37008 Salamanca, Spain
[2]School of ECE, College of Engineering, University of Tehran, Tehran, Iran
[3]Department of Electrical and Electronic Engineering, Imperial College, London, UK.
[4]Research Institute of Electrical Communication, Tohoku University, Sendai, Japan

Email: juanandn@usal.es

Abstract. We report on experimental and theoretical studies of nanoscale gate-lengths strained Silicon MODFETs as room temperature non resonant detectors. Devices were excited at room temperature by an electronic source at 150 and 300 GHz to characterize their sub-THz response. The maximum of the photovoltaic response was obtained when the FET gate was biased at a value close to the threshold voltage. Simulations based on a bi-dimensional hydrodynamic model for the charge transport coupled to a Poisson equation solver were performed by using Synopsys TCAD. A charge boundary condition for the floating drain contact was implemented to obtain the photovoltaic response. Results from numerical simulations are in agreement with experimental ones. To understand the coupling between terahertz radiation and devices, the devices were rotated at different angles under excitation at both sub-terahertz frequencies and their response measured. Both NEP (Noise Equivalent Power) and Responsivity were calculated from measurements. To demonstrate their utility, devices were used as sensors in a terahertz imaging system for inspection of hidden objects at both frequencies.

1. Introduction

The development of novel materials, concepts and device designs for terahertz radiation detection using semiconductors has recently fuelled the research of room temperature THz detectors. In early 90's, Dyakonov et al.[1] theoretically demonstrated the possibility of using sub-micron field effect transistors (FETs) as detectors of terahertz radiation by means of the oscillations of plasma waves in their channel in frequency ranges far beyond the transistor cut-off frequency (f_T). FET devices present many advantages: low cost, small size, room temperature operation and high-speed response that make them highly competitive with other THz detection technologies.

Room temperature detection of sub-terahertz radiation has been demonstrated by using different types of transistors such as commercial GaAs FETs and Si-MOSFET. In the present work, we investigated room temperature terahertz detection using Strained-Silicon Modulation Field Effect transistors (s-Si MODFET) with different gate lengths. As Schottky gated s-Si MODFETs exhibit higher values of channel electron mobility ($\approx$ 1400 $cm^2V^{-1}s^{-1}$) than conventional Si-MOSFET ($\approx$ 200 $cm^2V^{-1}s^{-1}$), they are excellent candidates to detect THz radiation, while offering many of the advantages of Si devices. The natural substrate compatibility of s-Si MODFETs with conventional Si technologies is a key benefit of the system Si:SiGe to compete in future on-chip THz system integration.

2. Device and TCAD Modeling

In this work, we studied n-channel s-Si/SiGe MODFETs as sub-THz detectors. The material system Si/SiGe allows the creation of a thin layer of strained silicon under tetragonal (biaxial tensile) strain. This strain has significant impact on the FET channel mobility of the semiconductor. The epistructure of the MODFETs used in this work is described in [2] and it is given in Fig. 1(a). All devices had a T-shape design [Fig. 1(b)] and were mounted on the same dual inline package (DIL14).

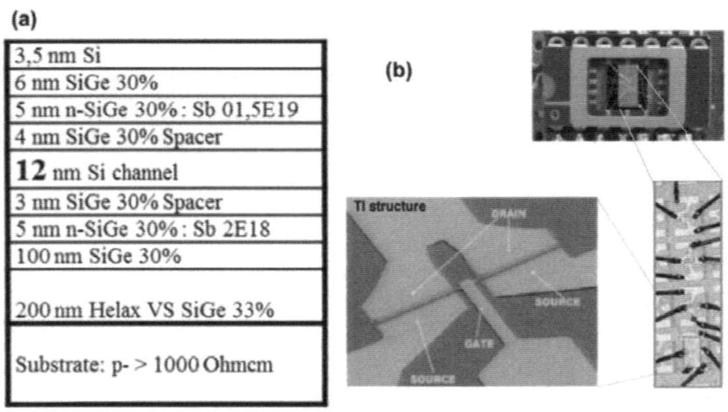

Figure 1. (a) Epi-structure of devices (b) T-shape design and devices on DIL14 with a microscope image of the devices.

Table 1 shows the geometrical parameters of the three set of transistors studied; a typical transfer characteristic is presented in the inset of Fig. 2(a).

				150 GHz		300 GHz	
	L_{SD} (μm)	L_g (μm)	W_{SD} (μm)	NEP ($pW/Hz^{0.5}$)	Rv (V/W)	NEP ($pW/Hz^{0.5}$)	Rv (V/W)
Device 1	2	100	80	158	~10	520	3
Device 2	2	150	80	125	15	546	~4
Device 3	2	500	80	400	~10	430	~10

Table 1. Geometrical parameters of the Si MODFETs along with measured and calculated NEPs and Rv.

A 2D numerical study was performed to understand and predict the response of the devices. Owing to the nature of the carrier transport in submicron length gate devices, we used a hydrodynamic (2DHD) model for the majority carriers (electrons) self-consistently coupled to a two-dimensional solution of the Poisson equation. The study of the THz photovoltaic response of the transistors was implemented in Synospys TCAD, as in measurements: the source was grounded and the gate was DC-biased, while floating the drain contact. A full description of the methods used for calculations could be found in [3].

3. Results and discussions

The devices were excited at room temperature by a dual-frequency electronic source based on frequency multipliers. The source frequencies were 150 and 300 GHz with output power levels of 3 mW and 6 mW respectively. The incoming radiation was modulated by a mechanical chopper between

233 Hz and 1.29 KHz. Induced photoresponse was measured by using the lock-in technique. Typical obtained photoresponses signals vs gate voltage are shown in Fig. 2 (a) for device 1 with Lg = 100 nm under excitation of 150 GHz (blue square symbols) and 300 GHz (blue dotted symbols). Figure 2 (b) gives the simulated photovoltaic response of the same device as in Fig. 2(a) when it was excited at several frequencies in the range 0.1-1THz. The shapes of both signals are similar and maxima are obtained around the devices' threshold voltages. This behavior is explained as non-resonant (broadband) detection that must be attributed to low values of the quality factor ($Q=\omega\tau<1$). Quality factor was estimated to have a value close to 0.13 at 150 GHz and 0.27 at 300 GHz.

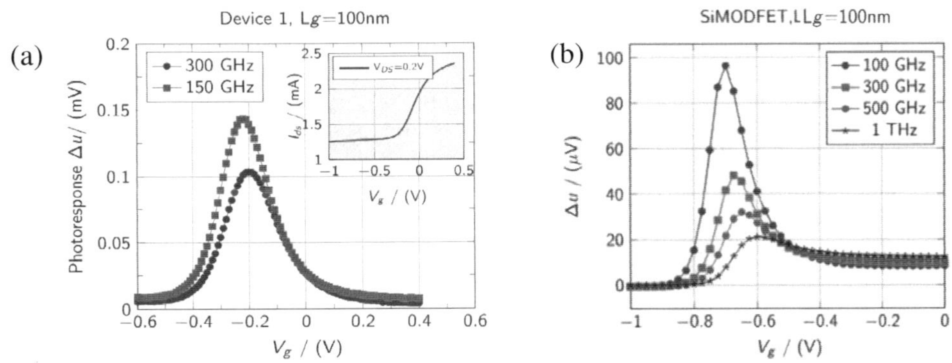

Figure 2. (a) Photoresponse vs. gate voltage under excitation at 150 GHz (red square) and 300 GHz (blue dotted). Inset shows the corresponding transfer characteristics. (b) Simulated photoresponse vs. gate voltage under excitation at different frequencies for Device 1 with Lg = 100 nm.

In order to understand how radiation is coupled to the devices we studied the photoresponse as a function of the polarization of the incoming radiation. It is clearly seen at low frequencies (150 GHz) bonding wires and contact pads could play the antenna role showing the maximum response at different angles (Fig (3). We observed that for devices with L_g=100 and 150 nm, only two lobs were observed and for device with higher gate length (L_g=500nm) different maximum were observed at different angles. Since the area is much bigger for this device, the coupling could be performed by both the bonding wires as well as the gate pads fingers.

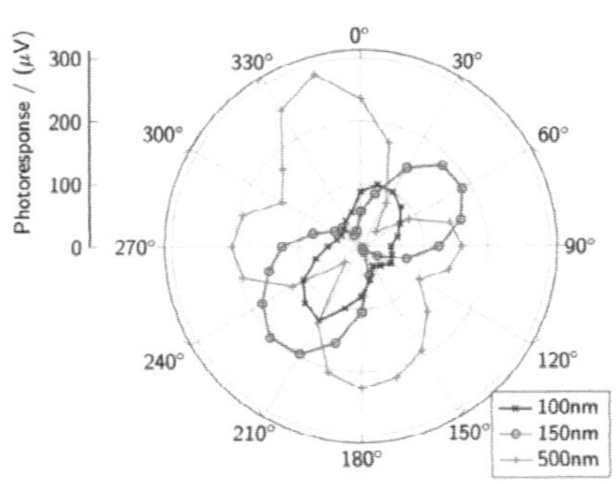

Figure 3. Photoresponse vs. rotation angle for all devices under excitation of 150 GHz.

Responsivity (R_v) as well as the Noise Equivalent Power (NEP) are the key parameters to determine the performance of these detectors' NEP and R_v are calculated by the formulas described in [4]. For the calculations the power incident at the detector position was 0.5/1 mW at 150/300 GHz. Table 1 summarizes the obtained NEP and R_v for the studied devices at 150 and 300 GHz. Device 2 with Lg = 150 GHz shows the best performance at 150 GHz. This must be attributed to a larger photoresponse signal exhibited by this device and to a better coupling with the incoming THz radiation. Those obtained values of NEP and R_v are comparable to the commercial room temperature THz detectors like Golay cells, pyroelectric detectors, and Schottky diodes. However, the Si-MODFET presents the advantage of being able to work at higher values of the modulation frequency (higher than 5 kHz).

To demonstrate its usefulness for practical applications, the detectors were used as the sensor within a THz imaging system. Figure 8 shows the THz images obtained at 150 GHz and 300 GHz as well as the visible one. Better resolution was obtained at 300 GHz which is related to its lower wavelength (λ=1mm).

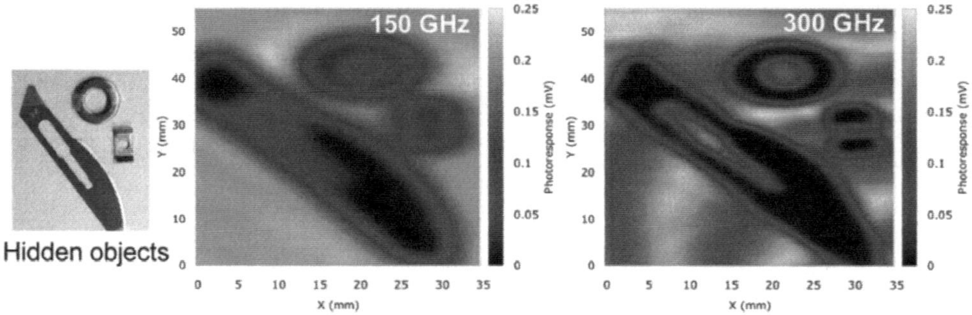

Figure 4. Visible image (left) and terahertz ones (right) at 150/300 GHz obtained at room temperature with Device 3 with Lg = 500 nm.

4. Conclusions

We report on a non-resonant detection of sub-THz radiation at 150 and 300 GHz at room temperature by using s-Si MODFETs with different gate lengths. Simulations results based on 2D numerical studies with Synopsys TCAD show a non-resonant response in agreement with experimental measurements. We have shown that gate pads are playing the role of antenna to couple the THz radiation at high frequencies. Competitive values of the performance parameters (NEP and R_v) were obtained. The practical use of these detectors for inspection of hidden objects was demonstrated.

ACKNOWLEDGEMENTS
This work was financially supported by the MINECO/FEDER Research Grant #TEC2015-65477-R and FEDER/Junta de Castilla y León Research Grant #SA045U16.

References

[1] Dyakonov M and Shur M 1996 IEEE Transactions on Electron Devices 43 380-7
[2] Y.M. Meziani , E. Garcìa-Garcìa, J.E. Velázquez-Pérez, D. Coquillat, N. Dyakonova, W. Knap, Grigelionis, K. Fobelets , Solid-State Electronics 83 (2013) 113–117
[3] J. A. Delgado Notario, E. Javadi1, J. Calvo-Gallego, E. Diez, J. E. Velázquez, Y. M. Meziani, K. Fobelets, Int. J. Hi. Spe. Ele. Syst. 25, 1640020 (2016)
[4] R. Tauk, F. Teppe, S. Boubanga, D. Coquillat, and W. Knap, Y. M. Meziani, C. Gallon, F. Boeuf, and T. Skotnicki, C. Fenouillet-Beranger, D. K. Maude, S. Rumyantseva and M. S. Shur, APPLIED PHYSICS LETTERS 89, 253511 2006

IOP Conf. Series: Journal of Physics: Conf. Series **906** (2017) 012004 doi:10.1088/1742-6596/906/1/012004

Inter-Layer Coupling Effects on Vertical Electron Transport in Multilayer Graphene Nanoribbons

F. Hashimoto and N. Mori

Division of Electrical, Electronic and Information Engineering, Osaka University, Osaka 565-0871, Japan

E-mail: hashimoto@si.eei.eng.osaka-u.ac.jp

Abstract. We have performed nonequilibrium Green function simulations on the vertical transport characteristics in multilayer graphene nanoribbons sandwiched between two graphene contact layers with varying the inter-layer coupling strength β. We find that the integrated transmission function thorough the top (or bottom) layer is hardly affected by β when the incident energy is close to the Dirac point. The β-insensitive energy window becomes narrower as the channel length increases.

1. Introduction

Graphene exhibits ultrahigh carrier mobility owing to the massless Dirac fermions and is expected to be used as a channel material of transistors [1–4]. In the transistor application, a multilayer graphene is more favorable for enhancing the magnitude of the ON current. However, a strong inter-layer coupling loses the linear band dispersion [5] and may cause a reduction of the carrier mobility. Recently, multilayer graphene with a turbostratic structure have been grown experimentally [6]. The transport properties are reported to be improved by the weak inter-layer coupling of the turbostratic structure. The mechanism of the improvement is, however, not well understood yet. Here we calculate vertical transport characteristics in multilayer graphene nanoribbons (GNRs) sandwiched between two graphene contact layers using the nonequilibrium Green function (NEGF) method with varying the inter-layer coupling strength β.

2. Calculation Method

We consider a device model whose schematic diagram is given in Fig. 1. The channel consists of AB-stacked N-layer GNRs structure of length L along the x-direction and of infinite length along the y-direction. We assign an index l for each layer ($l = 1, 2, \ldots, N$). Each layer is coupled to the neighboring layers with the inter-layer transfer integral β, which can be different from the intra-layer transfer integral $t_\perp$ of graphite. Semi-infinite graphene electrodes are connected to the bottom ($l = 1$) and top ($l = N$) GNR layers. Therefore, the set-up clearly comprises one or more GNRs sandwiched between two graphene contact layers.

We use the NEGF formalism [7–9] combined with a nearest-neighbor tight-binding approximation to calculate integrated transmission functions $\tau_{\rm TL}(E)$ from the top-right electrode to the top-left electrode, $\tau_{\rm BL}(E)$ to the bottom-left electrode, and $\tau_{\rm BR}(E)$ to the bottom-right electrode under the ballistic condition (no scattering). Along the y-direction, we apply the periodic boundary condition and use the eigenstate representation labeled by k_y. For a channel

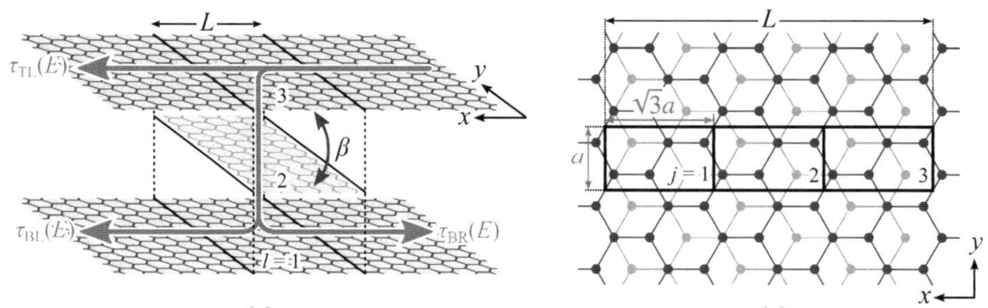

Figure 1. (a) Device model for the NEGF simulation. AB-stacked N-layer GNRs structure of length L is connected to semi-infinite graphene electrodes at the both ends of the bottom and top layers. We assign an index l for each layer ($l = 1, 2, \ldots, N$). Each layer is coupled to the neighboring layers with the inter-layer transfer integral β. (b) Top view of the multilayer GNR structure. The thick black rectangles define the unit cells. We assign an index j for each unit cell ($j = 1, 2, \ldots, M$). These figures illustrate the case for $N = 3$ and $M = 3$

consisting of M unit cells of the channel length $L = M \times \sqrt{3}a$ (a is the lattice constant), the total device Hamiltonian matrix, whose size is $(4MN) \times (4MN)$, can be written as

$$\hat{\mathcal{H}}_{\text{dev}} = \begin{bmatrix} H_1 & S & & \\ S^\dagger & H_2 & S & \\ & & \ddots & \\ & & S^\dagger & H_M \end{bmatrix}. \tag{1}$$

Here H_j is the Hamiltonian matrix of the size $(4N) \times (4N)$ of an isolated unit cell j, and S represents the transfer between the neighboring unit cells. H_j and S can be written as

$$H_j = \begin{bmatrix} h_1 & s_{12} & & \\ s_{12}^\dagger & h_2 & s_{23} & \\ & & \ddots & \\ & & s_{N-1,N}^\dagger & h_N \end{bmatrix}, \quad S = \begin{bmatrix} s_1 & & & \\ & s_2 & & \\ & & \ddots & \\ & & & s_N \end{bmatrix}. \tag{2}$$

Here h_l is the Hamiltonian matrix of the size 4×4 for the 4 atoms of the l-th layer in the j-th unit cell, and $s_{ll'}$ represents the inter-layer transfer between the layers l and l' in the j-th unit cell. s_l describes the transfer between the neighboring unit cells in the layer l.

The effects of the electrodes are accounted for by self-energy matrices; $\hat{\Sigma}_{\text{TR}}(E)$ is the self-energy matrix of the size $(4MN) \times (4MN)$ for the top-right, $\hat{\Sigma}_{\text{TL}}(E)$ the top-left, $\hat{\Sigma}_{\text{BR}}(E)$ the bottom-right, and $\hat{\Sigma}_{\text{BL}}(E)$ the bottom-left electrode. They are given by

$$\hat{\Sigma}_{p\text{R}}(E) = \begin{bmatrix} 0 & & & \\ & \ddots & & \\ & & 0 & \\ & & & \Sigma_{p\text{R}}(E) \end{bmatrix}, \quad \hat{\Sigma}_{p\text{L}}(E) = \begin{bmatrix} \Sigma_{p\text{L}}(E) & & & \\ & 0 & & \\ & & \ddots & \\ & & & 0 \end{bmatrix}, \quad (p = \text{T, B}), \tag{3}$$

where $\Sigma_{p\text{R}}(E)$ and $\Sigma_{p\text{L}}(E)$, whose sizes are $(4N) \times (4N)$, are given by

$$\Sigma_{\text{T}q}(E) = \begin{bmatrix} 0 & & & \\ & \ddots & & \\ & & 0 & \\ & & & \sigma_{\text{T}q}(E) \end{bmatrix}, \quad \Sigma_{\text{B}q}(E) = \begin{bmatrix} \sigma_{\text{B}q}(E) & & & \\ & 0 & & \\ & & \ddots & \\ & & & 0 \end{bmatrix}, \quad (q = \text{R, L}). \tag{4}$$

Here $\sigma_{\rm TR}(E)$ is the self-energy matrix of the size 4×4 for the top-right, $\sigma_{\rm TL}(E)$ the top-left, $\sigma_{\rm BR}(E)$ the bottom-right, and $\sigma_{\rm BL}(E)$ the bottom-left electrode. These self-energies are calculated using the method of Ref. 10.

In the ballistic limit, the integrated transmission functions from the top-right electrode can be written as

$$\tau_m(E) = \frac{1}{2\pi}\int \text{trace}[\Gamma_m(E)G(E)\Gamma_{\rm TR}(E)G^\dagger(E)]dk_y,\ (m = \text{TL, BR, BL}). \quad (5)$$

with $G(E) = [E - \hat{\mathcal{H}}_{\rm dev} - \hat{\Sigma}_{\rm TR}(E) - \hat{\Sigma}_{\rm TL}(E) - \hat{\Sigma}_{\rm BR}(E) - \hat{\Sigma}_{\rm BL}(E)]^{-1}$, $\Gamma_m(E) = i[\hat{\Sigma}_m(E) - \hat{\Sigma}^\dagger_m(E)]$ (m = TR, TL, BR, BL). We adopt the recursive Green function algorithm [11] for calculating the Green function $G(E)$.

To investigate the inter-layer coupling effects on the vertical transport characteristics, we calculate the integrated transmission functions with varying the inter-layer transfer integral, β, from 0 to $t_\perp$ (= 0.39 eV) of graphite [3].

3. Results and Discussion

Figure 2 shows $\tau_{\rm TL}(E)$ and $\tau_{\rm BL}(E)$ of a multilayer GNR with $N = 3$ and $L = 10$ nm for $\beta = 0$, 0.05, 0.1, and 0.3 eV. $\tau_{\rm BR}(E)$ (not shown in Fig. 2) exhibits similar behavior to $\tau_{\rm BL}(E)$. $\tau_{\rm TL}(E)$ for a monolayer graphene ($\beta = 0$) have a linear relation near the Dirac point as shown by dashed lines in Fig. 2. As β increases $\tau_{\rm TL}(E)$ decreases except for an energy region close to the Dirac point, where $\tau_{\rm TL}(E)$ are found to be hardly affected by β. This is because the GNR layers sandwiched between the bottom and top graphene layers open a bandgap, which prevents vertical electrons' flow between the top and bottom layers near the Dirac point. Figure 3 shows $\tau_{\rm TL}(E)$ of a multilayer GNR with $L = 10$ nm and $\beta = 0.05$ eV for $N = 2$, 3, 4, and 5. We see the β-insensitive energy window near the Dirac point regardless of N except for $N = 2$ where β affects the transmission even close to the Dirac point. This is due to the absence of nanoribbon layers for $N = 2$.

Figure 4 shows $\tau_{\rm TL}(E)$ of a multilayer GNR with $N = 3$ and $\beta = 0.05$ eV for $L = 10$, 15, and 20 nm. We find that the β-insensitive energy window becomes narrower as the channel length, L, increases. We introduce the critical inter-layer coupling, $\beta_{\rm c}$, which is defined by β for which $\tau_{\rm TL}(k_{\rm B}T)$ ($T = 300$ K) becomes 95% of that for $\beta = 0$ (see dotted line in Fig. 4). Figure 5 shows

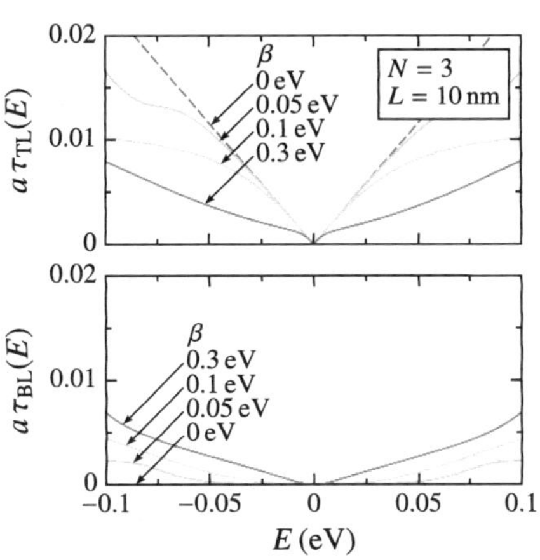

Figure 2. $\tau_{\rm TL}(E)$ and $\tau_{\rm BL}(E)$ for $\beta = 0$, 0.05, 0.1, and 0.3 eV. Red dash straight lines are $\tau_{\rm TL}(E)$ for a monolayer graphene.

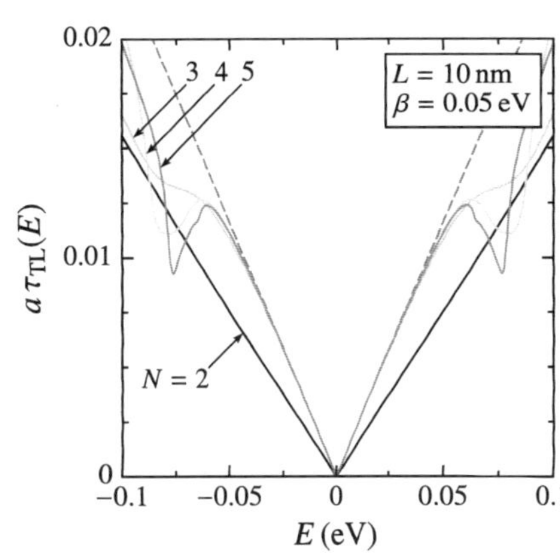

Figure 3. $\tau_{\rm TL}(E)$ for $N = 2$, 3, 4, and 5. Red dash straight lines are $\tau_{\rm TL}(E)$ for a monolayer graphene.

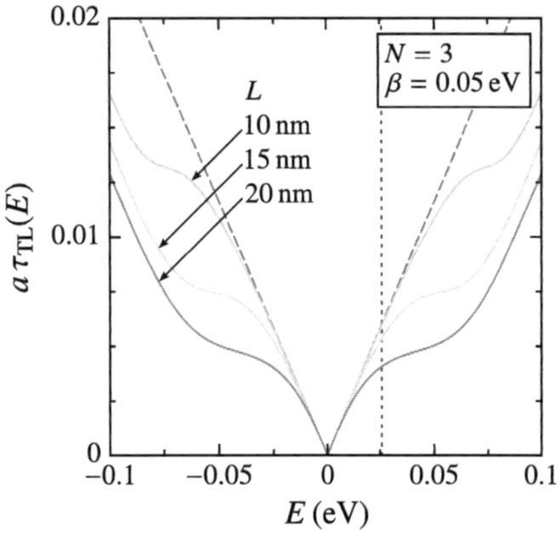

Figure 4. $\tau_{\mathrm{TL}}(E)$ of $L = 10$, 15, and 20 nm for $\beta = 0.05$ eV. Red dashed lines show $\tau_{\mathrm{TL}}(E)$ for a monolayer graphene. Blue dotted line represents the thermal energy at 300 K.

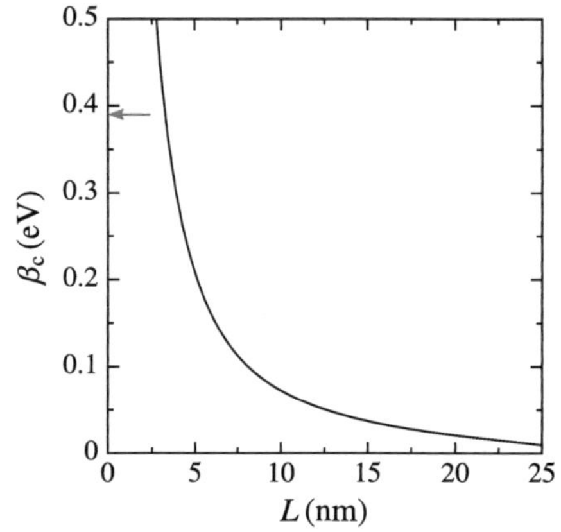

Figure 5. Channel length, L, dependence of critical inter-layer coupling strength β_c. Red arrow indicates $\beta = t_\perp$ of graphite.

channel length, L, dependence of β_c. For a smaller β, the channel length can be increased while keeping the β-insensitive energy window.

4. Conclusion

We have calculated vertical transport characteristics in multilayer GNRs sandwiched between two graphene contact layers by using NEGF method combined with a tight-binding approximation with varying the inter-layer coupling strength β. We find that the integrated transmission function thorough the top (or bottom) layer of the multilayer GNR is hardly affected by the inter-layer coupling strength when the incident energy is close to the Dirac point. The β-insensitive energy window becomes narrower as the channel length increases.

Acknowledgment

This work was supported by MEXT Grant-in-Aid for Scientific Research on Innovative Areas "Science of hybrid quantum systems" (Grant No. 2703).

References

[1] Geim A K and Novoselov K S 2007 *Nature Mater.* **6** 183
[2] Geim A K 2009 *Science* **324** 1530
[3] Neto A H C, Guinea F, Peres N M R, Novoselov K S, and Geim A K 2009 *Rev. Mod. Phys.* **81** 109
[4] Schwierz F 2010 *Nature Nanotechnol.* **5** 487
[5] McCann E and Koshino M 2013 *Rep. Prog. Phys.* **76** 056503
[6] Negishi R, Ohno Y, Maehashi K, Matsumoto K, and Kobayashi Y 2012 *Jpn. J. Appl. Phys.* **51** 06FD03
[7] Datta S 1995 *Electronic Transport in Mesoscopic Systems* (Cambridge: Cambridge University Press)
[8] Haug H and Jauho A P 1996 *Quantum Kinetics in Transport and Optics of Semiconductors* (Berlin: Springer)
[9] Mori N, Edagawa T, Kamakura Y, and Eaves L 2014 *Jpn. J. Appl. Phys.* **53** 04EN04
[10] Luisier M, Schenk A, Fichtner W, and Klimeck G 2006 *Phys. Rev. B* **74** 205323
[11] Lake R, Klimeck G, Bowen C R, and Jovanovic D 1997 *J. Appl. Phys.* **81** 7845

IOP Conf. Series: Journal of Physics: Conf. Series **906** (2017) 012005 doi:10.1088/1742-6596/906/1/012005

Transport Scaling Limits of Ovonic Devices: a Simulative Approach

C. Jacoboni[1], E. Piccinini, R. Brunetti[1], and M. Rudan

DEI Department and ARCES Research Center, University of Bologna, Viale Risorgimento 2, I-40136, Bologna, Italy
[1]FIM Department, University of Modena and Reggio Emilia, Via Campi 213/A, I-41125 Modena, Italy

E-mail: carlo.jacoboni@unimore.it

Abstract. The transport scaling limits of Ovonic devices are studied by means of a numerical solution of a time- and space-dependent transport model based on a set of equations that provide a good physical grasp of the microscopic process in hand. The predictivity of the approach has been confirmed through the comparison with recent experimental results where the parasitic effects have been reduced by the use of top-technology measuring equipments. The present analysis is performed for the AgInSbTe chalcogenide, since this material exhibits a steep threshold-switching dynamics which makes it promising for high-speed, non-volatile memory applications.

1. Introduction

The Ovonic threshold switch (OTS) is a characteristic electrical feature of many chalcogenide semiconductor materials. The process is very complex and still under analysis after about fifty years from its discovery [1]. Charge transport in Ovonic semiconductor materials is heavily influenced by the presence of trap states in the energy gap, which is confirmed by many structural analyses. This fact supports the applicability of a trap-limited transport scheme [2]–[7]. Recent experiments, performed with advanced testing equipments, focused on time-resolved electric response of nanometer chalcogenide layers [8], [9]. In fact, OTS has recently been exploited by the electronic industry as key physical mechanism of high-speed nano selectors for non-volatile memory (NVM) arrays [10]–[12]. Presently the scaling down of NVMs is reaching the 10-nm size [12], and the single memory selectors are scaled accordingly. Many experimental features can consistently be interpreted assuming that hot-carrier phenomena are dominant in the operating range of testing experiments and foreseen technological applications [7], [13].

Recently some Authors [14] suggested that scaling below 4 nm could induce structural changes in the electron states, thus spoiling the electric response of the device. In the present paper, we show that scaling below about 5 nm can prevent the setup of the microscopic non-equilibrium conditions that produce the electric switch. Our results were obtained using a numerical solution of a time- and space-dependent model described in [13] with the purpose of analysing transient transport regimes in Ovonic materials for the case of spatially non-homogenous electric fields. Here we focus our study on AgInSbTe (AIST) devices in view of the fact that AIST exhibits high-speed Ovonic switch when parasitic effects are engineered down to their minimum.

2. Theoretical framework and computational approach

The transport framework used in the present analysis is briefly summarized in this section; the details are provided in [13]. The model is of the hydrodynamic type and considers two levels for the carriers with energy difference Δ; the lowest level refers to trapped electrons, while the upper level refers to electrons in the band with mobility μ. Furthermore, a Maxwellian distribution of carriers at temperature T_e is assumed. Three equations are included for the particle continuity, local particle redistribution, and power balance, namely,

$$\frac{\partial n}{\partial t} = -\frac{\partial j}{\partial x}, \qquad \frac{\partial n_B}{\partial t} = \frac{\partial n}{\partial t} - \frac{n_B - \tilde{n}_B}{\tau_n}, \qquad n\,\frac{k_B\,T_e - k_B\,T_\circ}{\tau_T} = q\,j\,F + \Delta\,\frac{n_B - \tilde{n}_B}{\tau_n}, \quad (1)$$

where k_B is the Boltzmann constant, n_B and j the concentration and flux of carriers, $T_\circ$ the equilibrium temperature, n the total concentration (sum of band and trap concentrations), F the electric field, τ_n the recombination lifetime, and τ_T the temperature-relaxation time. Furthermore, $\tilde{n}_B$ is "tendential value" of the mobile-carrier concentration after a time variation of the electric field, and the following equality holds: $n/\tilde{n}_B = 1 + (g_T/g_B)\,\exp[\Delta'/(k_B\,T_e)]$, with g_T (g_B) the density of states of the traps (band), where the modified exponent $\Delta' = \Delta - \gamma\,|F|$ accounts for the fact that the excitation energy necessary to reach the band level is reduced by the field according to the Poole model. The model assumes that neither the lattice temperature nor the relaxation time τ_T vary within the operating time scale of the device [7]; finally, the electric field is obtained from the Poisson equation

$$\frac{\partial F}{\partial x} = \frac{q}{\epsilon}\,(n_\circ - n), \quad (2)$$

with ϵ the material permittivity and $n_\circ$ the background charge.

The model's equations are solved numerically for the unknowns n, F, n_B, and T_e, assuming the following constraints at $t = 0$: $F = 0$, $n(x, 0) = n_\circ$, and $T_e(x, 0) = T_\circ$ anywhere in the device. Furthermore, $n(0, t) = n(L, t) = n_\circ$ (with L the device length) at any time t; this implies $\partial n(x, t)/\partial t = 0$ and, from the Poisson equation, $\partial F(x, t)/\partial x = 0$ at $x = 0$ and $x = L$. Finally, the integral condition $V(t) = -\int_0^L F(x, t)\,\mathrm{d}x$ is imposed on F by the applied bias $V(t)$. The sensitivity of the results obtained with the above-described approach on both the physical parameters of the Ovonic material at hand and the device length have been reported in [13] for a "reference" case. The sensitivity on the crystal temperature has been analysed in [7] for the homogeneous case.

3. Results

The set of the four equations in (1)–(2) has been solved focusing on the AIST chalcogenide. The model has initially been fitted to the experimental data of [9], where the dynamics of the threshold switching is investigated for a 80 nm-long device at the picosecond time scale, and the effect of the measuring circuit is reduced to the minimum allowed by the present electronic technology. The parameters used for the calculations are listed in table 1, and the results are reported in figure 1. The model well interprets the results up to the threshold voltage. Just above threshold AIST crystallizes [9] (see the upper kink in the $j(V)$ curve), and the model cannot be applied since no thermal effects are included at present.

The potential technological interest of AIST is related to the rapid OTS switching (sub 50 ps delay time for the device tested in [9]), and the effect of the device length on the transport features is quite a relevant issue for the exploitation of this material for nanoelectronic applications. The present theoretical framework allows for a space-dependent analysis of the transport process, and this has been carried out here for three devices of 20, 40 and 80 nm. Some results enlightening the role of carrier heating on the onset of OTS are reported in figures 2 and 3.

IOP Conf. Series: Journal of Physics: Conf. Series **906** (2017) 012005 doi:10.1088/1742-6596/906/1/012005

Table 1. Parameters used for fitting the AIST current vs. voltage characteristic.

Symbol	*Definition*	*Value*	*Units*
Δ	Energy of mobile states from the trap level	0.315	eV
Γ	g_T/g_B	$1.4 \cdot 10^{-3}$	
γ	Coefficient of Poole effect	$5.0 \cdot 10^{-29}$	C m
ϵ	Relative permittivity	15	
μ	Mobility of mobile electrons	$6.0 \cdot 10^{-4}$	$m^2/(Vs)$
$n_\circ$	Equilibrium electron concentration	$1.45 \cdot 10^{27}$	m^{-3}
τ_T	Temperature relaxation time	$8.3 \cdot 10^{-13}$	s
τ_n	Recombination time	$1.0 \cdot 10^{-13}$	s
L	Device length	$80 \cdot 10^{-9}$	m
$T_\circ$	Room temperature	300	K

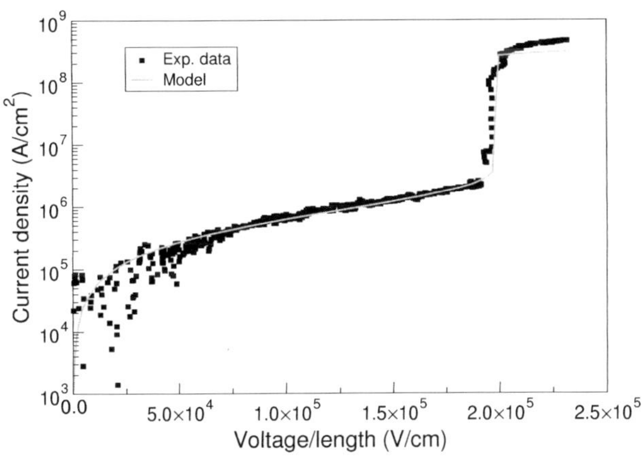

Figure 1. Experimental (black dots) and simulated (green line) current density vs. mean electric field for AIST. Experimental data are from [9].

Figure 2 confirms that the transport features of devices of different lengths depend only on the internal field and not on the applied voltage (see inset); in particular, the threshold voltage for the OTS in AIST devices is determined by a mean field of 2×10^5 V/cm. Finally, figure 3 shows the profiles of the internal field and relative concentration of mobile carriers at different voltages along the device for the $L = 20$ nm case. All the considered microscopic quantities show a significant variation within the first few nm from the injecting contact, then slowly reach constant values within 5 nm from the same contact. Below threshold the field is almost uniform in the device, while at and above threshold it reaches very high values just inside the device near the injecting contact. This process, fulfilling the energy balance included in the model, transfers enough energy to the electrons to sustain the necessary concentration of mobile carriers in the rest of the device. We can then conclude that the ultimate limitation to scaling, as far as the material properties are concerned, is the one related the the heating process. Scaling down to few nm is nowadays well outside the technological range. In order to go further down the nm scale, chalcogenides tailored to show more efficient heating effects must be explored.

4. Conclusions

The transient transport in OTS AIST-based devices has been analysed, assuming a trap-limited transport model, by means of a theoretical framework based on the solution of the three balance equations in (1), each encoding a basic physical principle, together with the Poisson equation (2). The limited computational burden of the model makes it possible to test a variety of conditions and, possibly, to incorporate it into more complex device-simulation tools.

A good comparison with available experiments is achieved for a 80 nm-long device. Moreover

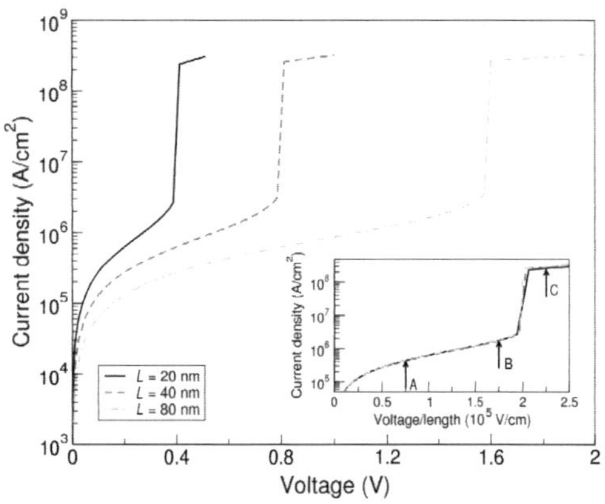

Figure 2. (top) $j(V)$ characteristic for AIST devices of different lengths. The inset shows the same result as a function of the mean field. A, B, and C show the cases considered in figure 3.

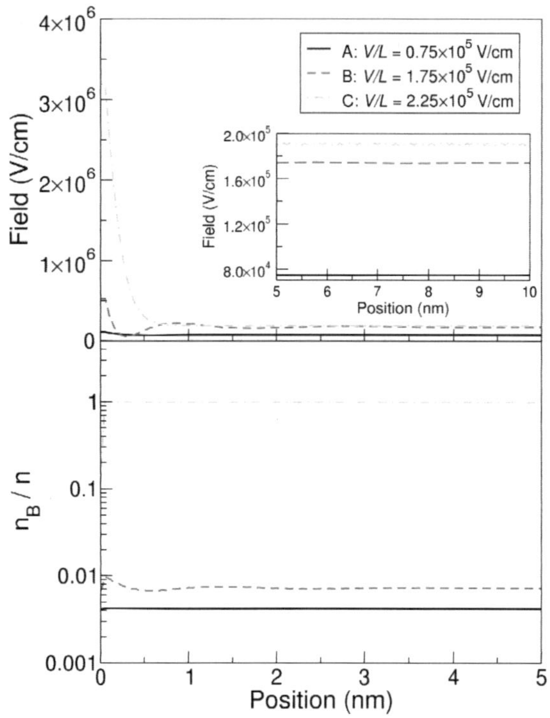

Figure 3. (right) Profiles of the electric field and fraction of electrons in the mobile states as functions of the position inside the device for a 20-nm layer of AIST at different values of the applied bias.

the analysis has been extended to shorter devices with the purpose of establishing the transport scaling limits of the OTS process in this material. As already obtained for other chalcogenides [13], the most significant spatial variability of the relevant quantities takes place within about few nm from the injecting contact. This suggests that the carrier heating process responsible for the OTS imposes a lower limit to the scaling down of OTS devices for nanoelectronic implementation.

References

[1] Ovshinsky SR 1968 *Phys. Rev. Lett* **22** 1450
[2] Ielmini D and Zhang Y 2007 *J. Appl. Phys.* **102** 054517
[3] Ielmini D 2008 *Phys. Rev. B* **78** 035308
[4] Piccinini E, Cappelli A, Buscemi F, Brunetti R, Ielmini D, Rudan M, and Jacoboni C 2012 *J. Appl. Phys.* **112** 083722
[5] Cappelli A, Piccinini E, Xiong F, Behnam A, Brunetti R, Rudan M, Pop E, and Jacoboni C 2013 *Appl. Phys. Lett.* **103** 083503
[6] Buscemi F, Piccinini E, Brunetti R, Rudan M, and Jacoboni C 2014 *Appl. Phys. Lett.* **104** 262106
[7] Piccinini E, Brunetti R, Bordone P, Rudan M, Jacoboni C 2016 *J. Phys. D: Appl. Phys.* **49** 495101
[8] Wimmer M and Salinga M 2014 *New J. Phys.* **16** 113044
[9] Shukla K D, Saxena N, Durai S, and Manivannan A 2016 *Sci. Rep.* **6** 37868
[10] Kau DC *et al.* 2009 *IEEE International Electron Devices Meeting*, p 617
[11] Anbarasu M, Wimmer M, Bruns G, Salinga M and Wuttig M 2012 *J. Appl. Phys.* **100** 143505
[12] Xiong F, Yalon E, Behnam A, Neumann C M, Grosse K L, Deshmukh S, Pop E 2016 *IEEE International Electron Devices Meeting* p 4.1.1
[13] Jacoboni C, Piccinini E, Brunetti R, Rudan M, 2017 *J. Phys. D: Appl. Phys.* **50** 255103
[14] Liu J and Anantram M P 2013 *J. Appl. Phys.* **113** 063711

IOP Conf. Series: Journal of Physics: Conf. Series **906** (2017) 012006 doi:10.1088/1742-6596/906/1/012006

Excitation polarization dependence of optical coupled exciton states in a single coupled quantum dot molecule

Heedae Kim[1*], Akihiro Murayama[1], Kwangseuk Kyhm[2], Jongsu Kim[3], Jindong Song[4], Koo Chul Je[5]

[1]Graduate School of Information Science and Technology, Hokkaido University, Japan
[2]Department of Opto-mechtronics and Cogno-mechatronics, RCDAMP, Pusan National University, Busan 609-735, Republic of Korea
[3]Department of Physics, Yeungnam University, Gyeongsan, 712-749, Republic of Korea
[4]Nano-Photonics Research Center, KIST, Seoul, 136-791, Republic of Korea
[5]College of Liberal Arts and Sciences, Anyang University, Gyeonggi-do, 430-714, Republic of Korea

*E-mail: hdkim@ist.hokudai.ac.jp

Abstract. The optically coupled exciton states have been investigated in terms of power dependent photoluminescence at the coupled GaAs quantum dots, which were laterally coupled along the $[1\bar{1}0]$ direction. The power dependent photoluminescence results are compared with two different excitation directions which are parallel to the direction of $[1\bar{1}0]$ and [110]. In coupled direction along the $[1\bar{1}0]$ direction, the power dependent redshift and coupled biexciton are observed, a fact that may be attributed to optical couplings via dipole-dipole interactions. Also, the redshift of excitons was manifested based on theoretical model.

1. Introduction

Recently, the droplet epitaxy techniques [1-3] allow possibly to fabricate laterally coupled quantum dot molecule (CQM). The vertically CQM structures have been intensively investigated in terms of optical coupling and entangling states [4-7]. The excitonic couplings were confirmed by electric fields manipulation [8,9], but optically induced couplings have not been considered so far. To control the excitonic states using electric fields, it is required to develop complicated fabrication process for making electrodes to apply for an external electric fields, but optically induced couplings can be conveniently studied without any fabrication processes. A conventional approach in coupling states was focused on vertically CQM, where interdot distance can be easily controlled. When the interdot distance is smaller than 10nm, the wavefunction overlaps between two dots are allowed by the tunneling effects. In the case of laterally CQM structures, the interdot distance is normally over 30nm which is too far to make tunneling effects directly. Here, we found that exciton-exciton interaction in a laterally CQM on excitation polarization via dipole-dipole interaction. When the excitation intensity of linearly polarized light parallel to the coupled direction of $[1\bar{1}0]$ is increased, excitons, local biexcitons, and a coupled biexciton of the two different quantum dots are observed with a power dependent redshift. To support the coupling states, the theoretical results by dipole-dipole interactions have been compared between different excitation laser polarization (parallel and perpendicular to the $[1\bar{1}0]$

IOP Conf. Series: Journal of Physics: Conf. Series **906** (2017) 012006 doi:10.1088/1742-6596/906/1/012006

direction of the sample). At the laser excitation parallel to the $[1\bar{1}0]$ direction, only clear redshift can be obtained theoretically as a result of dipole-dipole interactions.

2. Sample growth methods and experimental setups

The laterally CQMs were grown by VG80 solid-source molecular beam epitaxy (MBE) system on a GaAs (001) substrate. A 200nm thick GaAs buffer layer and a 100nm thick $Al_{0.3}Ga_{0.7}As$ barrier were grown at the substrate temperature (T_s) of 580°C. For the formation of Ga droplet, T_s was decreased to 300°C. The details of T_s calibration and chamber conditions for the DE were discussed [10,11]. The total amounts of Ga deposited on the $Al_{0.3}Ga_{0.7}As$ surface were 5 monolayer (ML) with a deposition rate of 0.1415 nm/s without arsenic supplying. For the fabrication of laterally CQMs, As_4 with a beam equivalent pressure of 1.5×10^{-5} Torr was supplied. The anisotropy of the Ga diffusion may be the major driving force for the formation of laterally CQMs along the direction of $[1\bar{1}0]$ on the (001) surface. After the formation of Ga droplets, T_s was subsequently decreased to 200°C for suppressing the Ga diffusion during the successive As supply. The photoluminescence (PL) of a laterally CQM was collected at 4K using a confocal arrangement, where frequency-doubled Ti:sapphire laser pulses (120fs pulse duration at a 80MHz repetition rate) were focused on the sample with a spot-size of 0.8 μm^2.

3. Results and Discussion

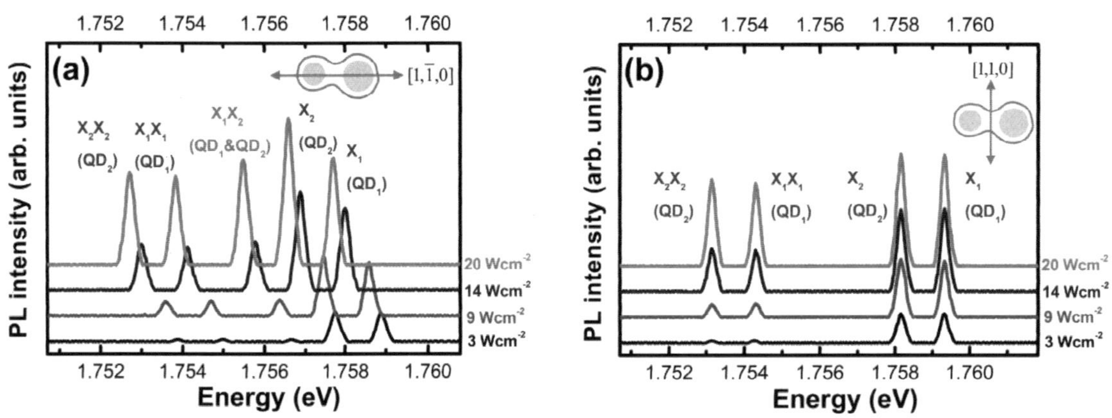

Figure 1. Experimental power dependent photoluminescence spectra in a single CQM. The excitation laser parallel to (a) $[1\bar{1}0]$ direction, and (b) [110] direction. In figure 1 (a), excitons, local biexcitons, and a coupled biexciton can be observed with power dependent redshift. (The red arrow line in the inset indicates the direction of laser polarizations and blue circles with black lines are corresponded to the GaAs QDs.)

A single laterally CQM was investigated by the power dependent PL measurements with different excitation polarizations (parallel and perpendicular to the $[1\bar{1}0]$ direction of the sample). The laser polarization directions are indicated in the inset of Figure 1 (a) and (b). The red arrow lines denote the direction of laser polarization on a laterally CQM, where blue solid circles with black lines are corresponded to the GaAs QDs. To investigate the optical anisotropy between $[1\bar{1}0]$ and [110] directions of the laterally CQMs, we performed the power dependent PL measurements with different laser polarization directions as shown Figure 1 (a) and (b). The PL spectra for the laser polarization parallel to $[1\bar{1}0]$ direction were significantly different from those for [110] direction as the excitation power densities are increased. Because the laterally CQMs were grown along the $[1\bar{1}0]$ direction, the laser polarization parallel to the $[1\bar{1}0]$ direction can trigger the optical couplings while no optical couplings are observed like a separate single QD in the laser polarization parallel to the [110] direction. In Figure 1 (a), when laser polarization was parallel to the $[1\bar{1}0]$ direction, the PL peak was significantly redshifted as increasing the excitation power. On the other hand, when the CQM was excited with the laser polarization parallel to the direction of [110] on increasing excitation power, the laterally CQM showed no redshift in PL peak even at 20 Wcm^{-2}. Furthermore, additional PL peak emerges at low energy (~ 1 meV) below with respect to the QD_2 compared to the PL peaks under [110] direction laser polarization in Figure 1 (b). This additional peak was characterized by a quadratic rise in its PL intensity and the faster decay time relative to the other two peaks (QD_1 and QD_2). Therefore, new additional peak can be attributed to a biexciton state where the two different exciton X_1 and X_2 originated from two different QDs as a result of optical couplings. Therefore, we can understand this biexciton state can be denoted like X_1X_2. These redshift and a coupled biexciton can be attributed from dipole-dipole interaction between two different QDs. Recently, it is reported that the dipole-dipole interaction is crucial as the dominant interaction mechanism in coupled QDs system [12]. The dipole-dipole interaction H_d is proportional to the dipole potential V_d ($V_d \propto d_a d_b/r^3$). Therefore, a sign of optical shift depends on the orientation alignments of two dipoles, where the strength is determined in terms of the so-called orientation factor (κ^2) [12-14].

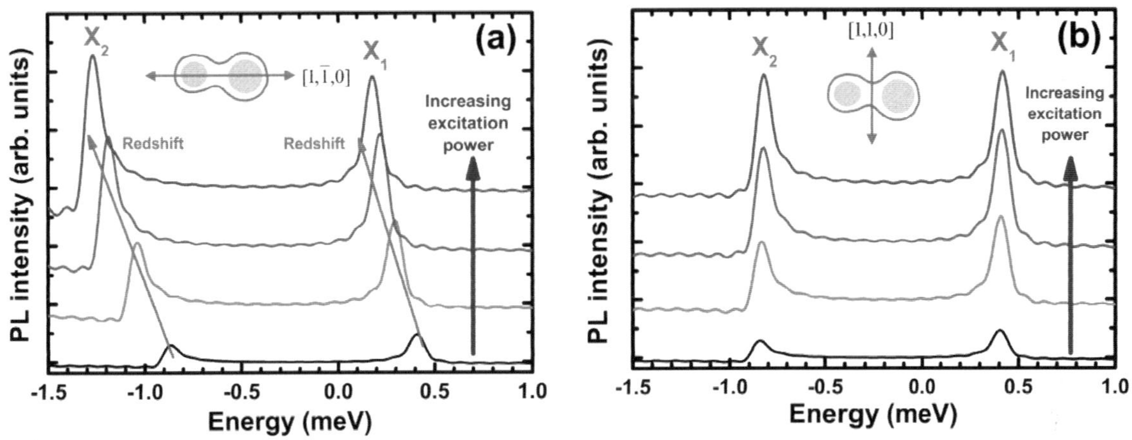

Figure 2. Power dependent photoluminescence spectra in a single CQM based on theoretical model. The excitation laser parallel to (a) $[1\bar{1}0]$ direction, and (b) [110] direction. In figure 2 (a), only clear redshift can be observed as a function of excitation power.

To understand the power dependent redshift with increasing the excitation power parallel to $[1\bar{1}0]$ direction, we performed the optical response theoretically in terms of the excitation-dependent absorption spectrum of a CQM with parallel dipoles from two different QDs. In Figure 2, the significant redshift can be obtained with increasing the excitation power parallel to the $[1\bar{1}0]$ direction, but we found no spectral shift with increasing the excitation power vertical to the $[1\bar{1}0]$ direction.

We consider a coupled two spherical quantum dots with a distance. The Hamiltonian of the coupled QDs is given by $H = H_{QD} + H_{dd}$, where H_{QD} is the optical excitations of excitons in the two QDs and H_{dd} is the exciton dipole-dipole interaction between two quantum dots.

$$H_{QD} = \sum_{\lambda_1} \hbar\omega_{\lambda_1} P_{\lambda_1\lambda_1} - E_0(t) \sum_{\lambda_1} \left(\mu_{\lambda_1} P_{\lambda_1 0} + \mu^*_{\lambda_1} P^*_{\lambda_1 0}\right) + \sum_{\lambda_2} \hbar\omega_{\lambda_2} P_{\lambda_2\lambda_2} - E_0(t) \sum_{\lambda_2} \left(\mu_{\lambda_2} P_{\lambda_2 0} + \mu^*_{\lambda_2} P^*_{\lambda_2 0}\right) ,$$

$$H_{dd} = \sum_{\lambda_1\lambda_1'\lambda_2\lambda_2'} \frac{e^2\rho_{\lambda_1\lambda_1'}\rho_{\lambda_2\lambda_2'}}{4\pi\varepsilon_0 R_{12}^3}\left(P_{\lambda_1\lambda_1'}P_{\lambda_2\lambda_2'}\right) = \sum_{\lambda_1,\lambda_2}(V_F\left(P_{0\lambda_1}P_{0\lambda_2}^* + P_{0\lambda_1}^*P_{0\lambda_2}\right) + V_D P_{\lambda_1\lambda_1}P_{\lambda_2\lambda_2})$$

where, $\lambda_{1(2)}$ indicates the eigen states of exciton states in $QD_{1(2)}$, $\hbar\omega_{\lambda_{1(2)}}$ is the corresponding eigen energy. We consider only the three exciton eigen states ($|\lambda> = |0>, |H>, |V>$), which are corresponded to the unexcited ground state and the parallel and vertical polarized excited states to the [1$\bar{1}$0] direction, respectively. The electric field E_0 (t) of a Gaussian laser pulse with 2.2 meV spectral width, and the interband optical transition matrix element $\mu_\lambda = <0|r|\lambda>$, and the operators $P_{\lambda\lambda'}$, which have the form of $|\lambda><\lambda'|$ in the bracket formation. As shown in Figure 2, dominant redshift appears with increasing the excitation power parallel to the [1$\bar{1}$0] direction, but no spectral shift is observed when the excitation power vertical to the [1$\bar{1}$0] direction is increased. In the case of the excitation power parallel to the [1$\bar{1}$0] direction, the coupling strength of the dipole-dipole interaction V_d^P = -1.24 meV is observed to be strong interaction strength compared to that of the direct Coulomb interaction V_c^P = +0.81 meV. On the other hand, when the excitation power vertical to the [1$\bar{1}$0] direction is applied, both of the coupling strength of the dipole-dipole interaction and direct Coulomb interaction are V_d^V = -0.35 meV and V_c^V = +0.40 meV, respectively. Consequently, the redshift with increasing excitation power parallel to the [1$\bar{1}$0] direction is investigated by dipole-dipole interaction. However, the redshift can be compensated by the blueshift due to the direct Coulomb interaction in the case of excitation vertical to the [1$\bar{1}$0] direction.

4. Conclusions

In conclusion, the presence of optically coupled exciton states in a long distance interdot CQM structures have been confirmed by the power dependent PL measurements which are compared with two different polarization parallel and vertical to the [1$\bar{1}$0] direction. In the case of the laser polarization parallel to the [1$\bar{1}$0] direction, the power dependent redshift and optical coupled biexciton state are experimentally observed and identified by a theoretical model based on dipole-dipole interactions. We have succeeded to demonstrate the coupling states can be controlled in terms of optical tuning directions.

References

[1] H. Kim, W. Lee, S. Park et al., 2017, *Sci. Rep.* **7**, 40026.
[2] H. Kim, R. Okuyama , K. Kyhm, M. Eto et al., 2016, *Nano Lett.* **16**, 27.
[3] H. D. Kim, K. Kyhm, R. A. Taylor, G. Nogues et al., 2013, Appl. Phys. Lett. **102**, 033112
[4] W. Sheng, and J. P. Leburton, 2002, *Phys. Rev. Lett.* **88**, 167401.
[5] K. M. Weiss, J. M. Elzerman, Y. L. Delley, J. Miguel Sanchez, and A. Imamoglu, 2012, *Phys. Rev. Lett.* **109,** 107401.
[6] J. R. Petta, A. C. Johnson, J. M. Taylor, E. A. Laird, A. Yacoby, M. D. Lukin, C. M. Marcus, M. P. Hanson, and A. C. Gossard, 2008, *Science* **320**, 772 .
[7] G. Bester, J. Shumway, A. Zunger, 2004, *Phys. Rev. Lett.* **93**, 047401.
[8] C. R. Kagan, and C. B. Murray, 2015, *Nat. Nanotechnol.* **10**, 1013.
[9] X. R. Zhou, J. H. Lee, G. J. Salamo, M. Royo et al., 2013, *Phys. Rev. B* **87**, 125309.
[10] M. Yamagiwa, T. Mano, T. Kuroda, T. Tateno, K. Sakoda, G. Kido, and N. Koguchi, 2006, *Appl. Phys. Lett.* **89**, 113115.
[11] J. S. Kim, M. S. Jeong, C. C. Byeon, D. K. Ko et al., 2006, *Appl. Phys. Lett.* **89**, 113115.
[12] H. Kim, I. Kim, K. Kyhm, R. A. Taylor et al., 2016, *Nano Lett.* **16**, 7755.
[13] J. R. Lakowicz, 2006, Principles of Fluorescence Spectroscopy (Springer, New York)
[14] H. Kim, and J. D. Song, 2018, *Appl. Surf. Sci.* **427**, 405-409

IOP Conf. Series: Journal of Physics: Conf. Series **906** (2017) 012007 doi:10.1088/1742-6596/906/1/012007

Full-Band modelling of phonons in polytype Ge and Si

J. Larroque, P. Dollfus, J. Saint-Martin

C2N, CNRS UMR 9001, Univ. Paris-Sud, Université Paris-Saclay, 91405 Orsay, France

Jerome.saint-martin@u-psud.fr

Abstract. The phonon dispersions and their related properties are computed in polytype materials by using a semi-empirical approach called adiabatic bond charge model. Both hexagonal 2H and cubic 3C phases of Silicon and Germanium are investigated in terms of heat capacity, Raman shift and sound velocities for each phonon branch in all main directions.

1. Introduction

Among the new nanostructures promising in terms of thermal engineering, polytype nanowires made of Silicon or Germanium [1,2] are particularly interesting. Indeed, in these nanomaterials, nano-clusters can crystallize in the 2H phase, although this phase cannot exist in the bulk material counterpart under normal conditions. These nano-clusters are embedded in standard 3C phase areas, which results in. structures exhibiting many interfaces that are quasi-periodically distributed at a scale that corresponds to the mean free path of phonons.

Our approach is based on a semi-empirical adiabatic bond charge model (ABCM) that is faster and easier to compute than ab-initio models, and even more accurate if using good fitting parameters. Due to its efficiency, the ABCM has been successfully used to calculate phonon dispersions in several materials [3] and in various silicon nanostructures including silicon nanowires [4] and nanocrystals [5,6].

In this paper, after reminding the principles of the ABCM calculation and of the interpolation method used to save memory resources, the resulting Full-Band (FB) dispersions are presented, i.e. for all modes and at all k-points in the Brillouin zone for both hexagonal 2H and cubic 3C phases in Si and Ge. Then, the related properties are discussed in terms of specific heat, Raman shift and group velocity.

2. Full-band phonon mode calculation using ABCM

Within the harmonic approximation and considering only small displacements of atoms $\vec{u}$, the equation of motion at the vibrational angular frequency ω can be written as, $\omega^2 \cdot \vec{u} = D \cdot \vec{u}$, where D is the dynamical matrix which contains all the relevant force fields.

To design an accurate dynamical matrix, the Weber's adiabatic bond charge model (ABCM) has been shown to give very good results for bulk diamond and zinc-blende-type crystals with only 4 empirical parameters [3]. The model is based on an atomistic description of the crystal in which 2 kinds of point-like charges are considered: those related to actual ions present in the unit cell and virtual particles called "Bond Charges" (BC). BC are located exactly in the middle of the segment between two neighbouring ions according to the adiabatic approximation. Weber's model consists in four different types of forces between charged particles: 2 central forces (one between ion cores –as in the force constant model- and another between ions and bond charges), 1 bond-bending interaction and, at last, coulomb interactions due to non-uniform charge distribution of the "bond charge" (derived via an Ewald transformation).This atomistic approach can be used to investigate the effect of a phase transformation in which the effect of the crystalline geometry prevails.

In this work, the cubic 3C (standard diamond phase) and hexagonal 2H phases of both Germanium and Silicon have been investigated by using the ABCM model. The values of the 4 semi-empirical parameters are those used in Ref. [3] for the Si and Ge in the cubic phase. The same parameters are also used to study the hexagonal phases. This approximation seems reasonable because in both phases each atom has 4 first nearest neighbours that are located in the vertex of a tetrahedron. Moreover, the distance between atoms is the same as shown experimentally in the case of 3C-Ge in [1] and theoretically in the case of 2H-Si in [7].

As previously done by [8] for electron dispersions to limit the memory consumption and reduce the computation time, an interpolation of phonon dispersions is used. That means that, only a limited number of dispersion values in the reciprocal space are stored. These points are located at the nodes of cubic components in which analytical interpolations are performed. Our study only involves cubic and hexagonal crystals and then the ensemble of (16x16x16) cubes that corresponds to an eighth of the BZ. The full description of the BZ is then obtained by using symmetry relationships and by knowing for each cube n and for each phonon mode m, 8 scalar parameters (vector) $\Omega_{0-7,m,n}$ ($\vec{V}_{0-7,m,n}$) for the interpolation of the pulsation $\omega_m(\vec{k})$ (the group velocity $\vec{v}_m(\vec{k})$, respectively). It yields for a given wave vector $\vec{k}(k_x, k_y, k_z)$:

$$\omega_m = \Omega_{0,m,n} + \Omega_{1,m,n}.k_x + \Omega_{2,m,n}.k_y + \Omega_{3,m,n}.k_z + \Omega_{4,m,n}.k_x k_y + \Omega_{5,m,n}.k_x k_z + \Omega_{6,m,n}.k_y k_z + \Omega_{7,m,n}.k_x k_y k_z$$

$$\vec{v}_m = \vec{V}_{0,m,n} + \vec{V}_{1,m,n}.k_x + \vec{V}_{2,m,n}.k_y + \vec{V}_{3,m,n}.k_z + \vec{V}_{4,m,n}.k_x k_y + \vec{V}_{5,m,n}.k_x k_z + \vec{V}_{6,m,n}.k_y k_z + \vec{V}_{7,m,n}.k_x k_y k_z$$

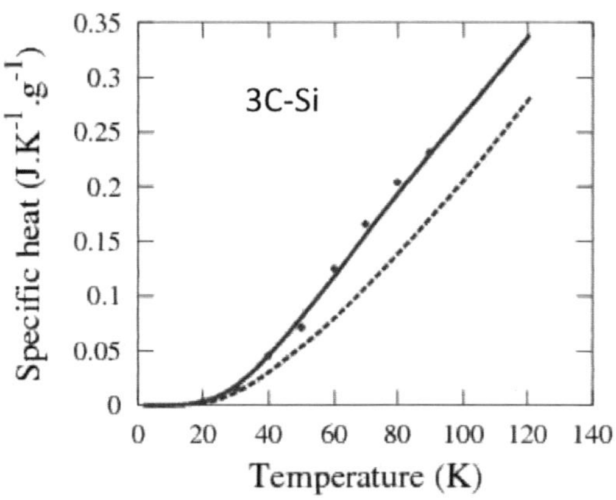

Fig. 1. Specific heat as a function of temperature in cubic Si: ABCM (Cont. line), isotropic model (dashed), experimental data [8] (symbols)

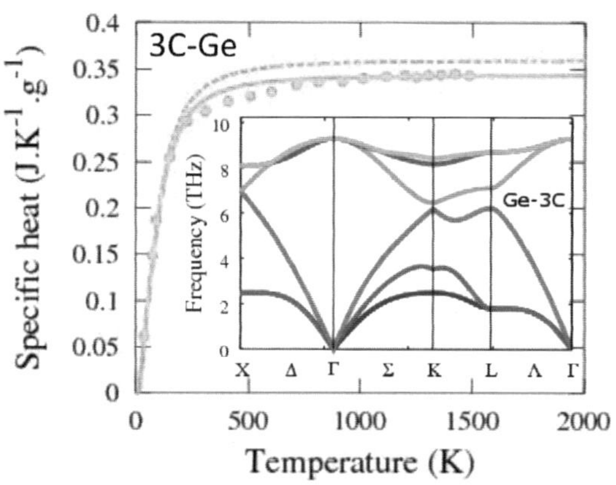

Fig. 2. Specific heat as a function of temperature in cubic Ge: ABCM (Cont. line), isotropic model (dashed), exp. data [9] (symbols). Inset: disperion.

3. Properties of cubic and hexagonal phases of Si and Ge

The specific heat has been computed for Si and Ge in the cubic 3C phase and is plotted in Fig. 1 and Fig. **2**, respectively. The phonon dispersion for Ge-3C is also plotted in the inset of Fig. **2**. These ABCM specific heats are compared with that computed from the frequently used quadratic isotropic approximation, i.e. $\omega_m(k) = a_m + b_m k + c_m k^2$, where a_m, b_m and c_m are the fitting parameters determined for the (100) direction [12-14]. The specific heat strongly depends on the density of states [11] and a full-band description is required to reproduce the correct temperature-evolution in both low and high temperature regimes. Indeed, with an appropriate normalization, the heat capacitance computed from an isotropic dispersion could fit experimental data at high temperature but would fail at

reproducing the correct evolution at low temperature or vice versa. Only the full-band description can accurately capture the specific heat in the full temperature range.
By using the same ABCM parameters than in the 3C case, the computed dispersions in hexagonal 2H crystals for both Si and Ge are plotted in Figure 3 a) and b). As expected, the density of states presented in Figure 3 c) for Ge is almost independent of the crystalline structure and in particular the total number of phonon modes, i.e. the integral of the DOS, are exactly the same for 3C and 2H phases, as they have the same mass density. Thus, the specific heat cannot be used to distinguish different phases, even at low temperature. However, the dispersion relationships are strongly different. Even along the most similar (main) directions, i.e. ΓX for 3C and ΓM for 2H, there is no obvious way to identify a correspondence between the twelve modes of Ge-2H (Fig. 3b) and the six modes of the Ge-3C (inset of Fig. 2b).

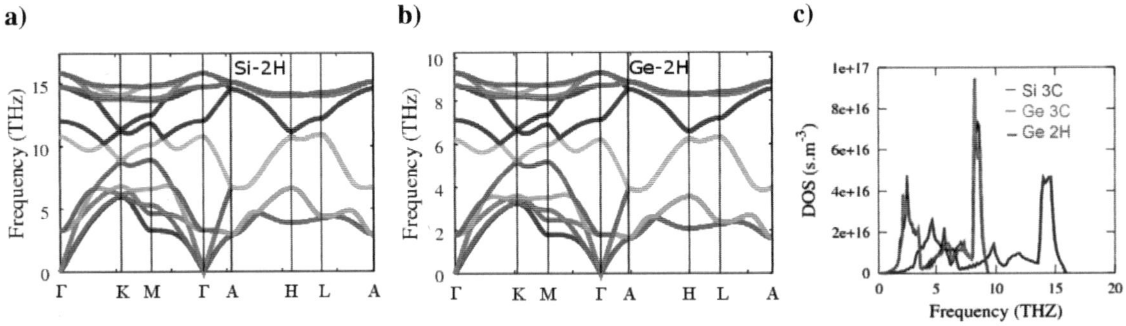

Figure 3. Phonon dispersion in a) hexagonal 2H Silicon, b) hexagonal 2H Germanium. c) Density of states for Si 3C (blue line), Ge 3C (green line) and Ge 2H (red line)

Table 1. Frequency of phonon in Γ for Ge 3C, Ge 2H, Si 3C and Si 2H.

		3C		2H					
Ge	Frequency (THz/cm^{-1})	0	9.29/310	0	1.79/60	6.21/207	7.12/237	8.70/290	9.30/310
	Degeneracy	3	3	3	2	1	1	2	2
Si	Frequency (THz/cm^{-1})	0	15.8/527	0	3.27/109	10.79/360	12.0/400	14.7/490	15.8/527
	Degeneracy	3	3	3	2	1	1	2	3

Table 2. Sound velocity (km/s) in Γ(ω~0)
in the main three directions for Ge 3C, Ge 2H, Si 3C and Si 2H.

material	Phase	direction	TA1	TA2	LA
Germanium	3C	[100]	347	347	497
		[110]	274	348	539
		[111]	300	300	555
	2H	[1-210]	301	317	535
		[10-10]	301	316	536
		[0001]	301	301	556
Silicon	3C	[100]	577	577	818
		[110]	459	577	890
		[111]	501	501	913
	2H	[1-210]	501	528	883
		[10-10]	501	528	883
		[0001]	501	501	914

In Table 1 are listed the values of phonon frequencies at the Γ point for both phases in Si and Ge. These quantities have been identified experimentally from Raman spectroscopy in Ref [2,5,15]. It can be

observed that the position of the highest optical branches in both 3C and 2H phases are the same. Besides, in Ge (Si) 2H phase, optical phonons at Γ have frequencies of 6.2 (10.8), 7.1 (12) or 8.7 (14.7) THz. Moreover, due to its highest level of degeneracy, the highest energy phonon branch should be more active in Raman experiment. In silicon 2H, the two highest Raman peaks experimentally measured in Ref [2] at 495.6 cm^{-1} and 515.2 cm^{-1}, due to TO folded modes, are present in ABCM calculations at 490 and 527 cm^{-1}, respectively. The slight discrepancy can be explained by the internal strain present in the measured Si structures.
In Table 2, the sound velocities for the (LA and TA) acoustic phonon branches (i.e. with $\omega = 0$ Thz at Γ) are reported for both 2H and 3C phases of Si and Ge. According to this classification of the lowest phonon branches, the sound velocity is very similar in both phases. It should be mentioned that in the standard 3C configuration, the transverse acoustic modes are degenerate only in the (100) direction. Besides, around the Γ point the degree of anisotropy of the acoustic dispersions, in terms of sound velocity, is higher than 10% in the cubic phase (for TA2), while it is lower than 5% in the hexagonal phase.

4. Conclusion

A semi-empirical ABCM model has been used to compute the FB phonon dispersion in both cubic and hexagonal phases of Si and Ge. The memory resources required to store the dispersions has been optimized thanks to an interpolation method. Due to the lack of experimental data on 2H-Si and 2H-Ge, we used the same empirical parameters in both 3C and 2H phases. Our results show that a FB approach is mandatory to capture the specific heat in a wide temperature range. The differences between the phases 3C and 2H are similar in Silicon and Germanium. The DOS, specific heat and sound velocity remain almost unchanged whatever the phase. Besides, our model is able to fit with Raman experiment results. Even if the bulk properties remain similar in almost all temperature ranges, the dispersion relationship are different between the 2 phases. This suggests that the transmission at polytype interfaces can lead to significant reduction of the transmission due to the phonon mismatch. To investigate this issue, a full-band transport model dedicated to structures with polytype interfaces is under development.

Acknowledgement

This work was supported by the Agence Nationale de la Recherche (ANR) through the project JCJC NOE ANR-12-JS03-0006 and by the IDEX Paris-Saclay under project ANR-11-IDEX-0003-02.

References

[1] Vincent L, Patriarche G et al. 2014 *Nano Lett.* **14** 4828–36
[2] Lopez F J, Givan U, Connell J G and Lauhon L J 2011 *ACS Nano* **5** 8958–66
[3] Weber W 1977 *Phys. Rev. B* **15** 4789
[4] Russell J P 1965 *Appl. Phys. Lett.* **6** 223
[5] Parker Jr J H, Feldman D W and Ashkin M 1967 *Phys. Rev.* **155** 712
[6] A. Valentin, J. Sée, S. Galdin-Retailleau, P. Dollfus 2008 *J. Phys.: Condens. Matter* **20** 145213
[7] Amato M, Kaewmaraya T, Zobelli A, Palummo M and Rurali R 2016 *Nano Lett.* **16** 5694–700
[8] Fischetti M V and Laux S E 1988 *Phys. Rev. B* **38** 9721
[9] Pearlman N and Keesom P H 1952 *Phys. Rev.* **88** 398
[10] Okhotin A S, Pushkarskij A S and Gorbachev V V 1972 (USSR: Atomizdat)
[11] Mounet N and Marzari N 2005 *Phys. Rev. B* **71** 205214
[12] Jean V, Fumeron S, Termentzidis K, et al. D 2014 *J. Appl. Phys.* **115** 024304
[13] Martin P N, Aksamija Z, Pop E and Ravaioli U 2010 *Nano Lett.* **10** 1120–4
[14] Pop E, Dutton R W and Goodson K E 2004 *J. Appl. Phys.* **96** 4998
[15] Russell J P 1965 *Appl. Phys. Lett.* **6** 223

Lateral interdot coupling among dense ensemble of InAs quantum dots grown on InP substrate observed at cryogenic temperatures

A. Maryński[1], M. Syperek[1], M. Pieczarka[1], M. Gawełczyk[1,2], J. Misiewicz[1], V. Liverini[3], M. Beck[3], J. Faist[3], and G. Sęk[1]

[1] Laboratory for Optical Spectroscopy of Nanostructures, Department of Experimental Physics, Faculty of Fundamental Problems of Technology, Wrocław University of Science and Technology, Wybrzeże Wyspiańskiego 27 27, 50-370 Wrocław, Poland

[2] Department of Theoretical Physics, Faculty of Fundamental Problems of Technology, Wrocław University of Science and Technology, Wybrzeże Wyspiańskiego 27 27, 50-370 Wrocław, Poland

[3] Institute of Quantum Electronics, ETH Zürich, Wolfgang-Pauli-Strasse 16, Zürich 8093, Switzerland

E-mail: `aleksander.marynski@pwr.edu.pl`

Abstract. The lateral interdot coupling is investigated in high density ($\sim 10^{11}$ cm^{-2}) self-assembled InAs quantum dots (QDs) grown on an InP substrate. Two types of structures are selected for this study, in which QDs are embedded into an InAlAs matrix, forming nearly twice stronger confinement for an electron and a hole than expected for an InAlGaAs counterpart. Resonantly injected low carrier population in these families of QDs gives very different spectral and temporal response in the temperature range of 5-30 K. While InAs/InAlGaAs QDs show monotonic temperature quench of photoluminescence (PL), the process seems to be ineffective in the family of InAs/InAlAs dots. Moreover, the PL decay traces for InAs/InAlGaAs QDs reveal a two-exponential decay as compared to a mono-exponential one observed for InAs/InAlAs dots. While a short decay component of $\leq$1.9 ns has been attributed to recombination of an electron-hole pair confined in the dot, the long one of >2.4 ns, observed exclusively for InAs/InGaAlAs QDs, is attributed to recombination of spatially separated electron-hole pairs formed due to carrier exchange between adjacent dots.

1. Introduction

The presence of lateral coupling among high surface density self-assembled quantum dots (QDs) in the absence of external electric field has been studied for many years, in the view of the appropriate description of the temperature activated carrier transfer process.[1, 2] At high temperatures, such processes can be well described with a carrier activation mechanism enhanced by a large density of phonon states that allows efficient carrier redistribution via higher order QDs states or the wetting layer (WL). Low-temperature processes, however, are still far for being well understood. Negligible phonon population at cryogenic temperatures makes above mentioned carrier transfer mechanism very inefficient. One can imagine that in such circumstances QD confined carriers can still be shared among closely spaced QDs either by: (i) resonant coupling of fundamental states, or (ii) coupling between the fundamental state in one dot to an excited state in the adjacent one. In both cases, the resonance condition is constrained to the thermal energy in the system.[3] In the ensemble of self-assembled QDs grown within the Stranski-Krastanow mode, the resonant coupling described in the first case is very unlikely, since local strain conditions prevent the nucleation of two identical dots in the close proximity that is an essential requirement to observe the effect. A very interesting and more probable is to observe the second process. First, more pronounced penetration of a carrier wave function into the barrier for an excited state of a dot increases the chance to be coupled to the adjacent dot. Second, among the inhomogeneous ensemble of dots one can find two non-identical QDs close to each other that can obey the second condition for coupling.

To look for the phenomenon of the lateral interdot carrier coupling at cryogenic temperatures the attention has been focused on InAs QDs grown on an InP(001) substrate, devoted for the laser applications in the third telecom spectral window (1.55 μm). The application target of these dots implicates their high surface density, reaching roughly 10^{11} cm^{-2}. Since the coupling condition, among many other parameters, depends on the barrier material, two structures are selected where QDs are embedded either in an InAlAs or InGaAlAs barrier. The first barrier provides nearly twice stronger carrier confinement in the dot than the second one. The experimental methodology is based on spectroscopic tools, including photoluminescence (PL), photoluminescence excitation (PLE), photoreflectance (PR), and time-resolved photoluminescence (TRPL).

2. Experimental details and methodology

Investigated QD structures were grown by molecular beam epitaxy on (001)InP substrates. Two structures were taken into account where InAs QDs formed by self assembly are embbeded into the $In_{0.52}Al_{0.48}As$ or $In_{0.53}Ga_{0.23}Al_{0.24}As$ matrix, lattice matched to InP. Slightly different thermodynamics of the QDs growth between two structures results in different dimensions and surface coverage. A typical in-plain size of InAs/InGaAlAs QDs is $\sim$12 nm$\times\sim$24 nm and the height is $\sim$3 nm, whereas InAs/InAlAs QDs are almost twice bigger in the plain, having the size of $\sim$40 nm$\times\sim$70 nm and $\sim$3 nm in height. While the InAs/InGaAlAs QDs have a surface density of $\sim 1\times10^{11}$ cm^{-2}, the InAs/InAlAs QDs have only slightly smaller density of $\sim 7\times10^{10}$ cm^{-2}. Further information about the growth conditions are in Ref.[4, 5].

For all the experiments, the QDs structures where held in an variable temperature He-flow cryostat, operating in the range of 5-300 K, with the accuracy of $\leq$1 K. For PL, TRPL, and PLE experiments, the structures were excited by a train of 160-fs long pulses, at a repetition frequency of 76 MHz, generated by an optical parametric oscillator with wavelength tunability between 1.1 μm and 1.6 μm. Subsequent PL emission was dispersed by a 0.3-m-focal-length monochromator and detected either by a liquid-nitrogen-cooled NIR-type streak camera or an InGaAs linear array detector. Details of the PR experiment are described in Ref.[6].

3. Results and discussion

Spectral response of investigated QD structures is presented in Fig. 1. The PR experiment revealed a spectral feature at $\sim$1.42 eV attributed to the interband absorption in the InP substrate. The second PR transition at $\sim$1.12 eV is observed only for the InAs/InGaAlAs QD structure and related to optical absorption in the $In_{0.53}Ga_{0.23}Al_{0.24}As$ matrix. Respective optical transition for the $In_{0}.52Al_{0.48}As$ matrix is expected at $\sim$1.56 eV (not shown). The absorption edge of the wetting layer is observed in the PLE experiments at $\sim$0.98 eV and $\sim$1.02 eV for QD structures with a quaternary and ternary barrier, respectively. The QD emission band is centered at $\sim$0.85 eV for InAs/InGaAlAs QDs, and at $\sim$0.81 eV for InAs/InAlAs QDs. Spectral broadening of the PL band (ΔE), obtained at the same excitation conditions, is $\sim$28 meV for QDs with a quaternary barrier, and is twice smaller than $\Delta E\approx56$ meV for QDs in the ternary matrix. Although ΔE generally reflects inhomogeneities within the QDs ensemble related to

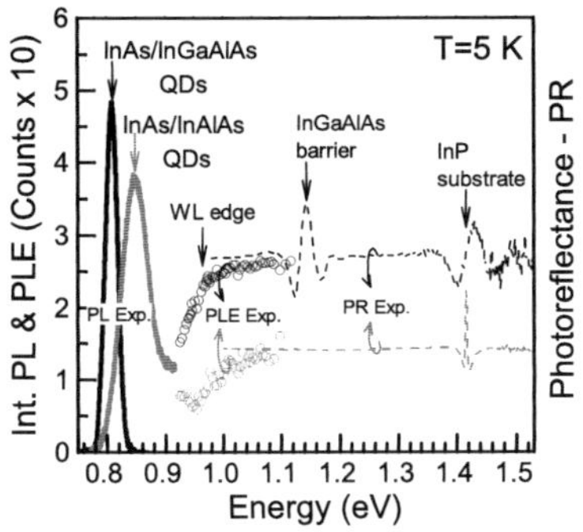

Figure 1. Photoluminescence, photoluminescence excitation, and photoreflectance spectra measured for InAs/$In_{0.53}Ga_{0.23}Al_{0.24}As$ and InAs/$In_{0.52}Al_{0.48}As$ QD structures. InP substrate, InGaAlAs barrier, and the wetting layer related optical transitions are indicated by arrows.

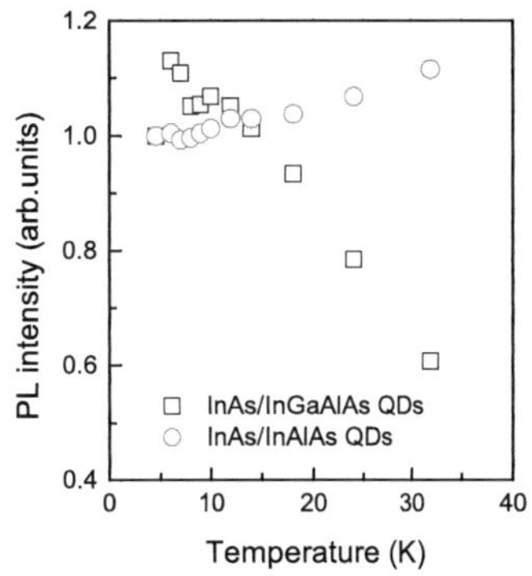

Figure 2. Photoluminescence quench registered for InAs/$In_{0.53}Ga_{0.23}Al_{0.24}As$ QDs (black squares) and InAs/$In_{0.52}Al_{0.48}As$ QDs (red circles) at the resonant excitation condition at 0.918 meV and photoexcitation of less than 1 e-h pair per QD.

strain, size and chemical content distribution, however, narrowing of the PL emission may also indicate a selection process due to the presence of interdot lateral carrier transfer.

To evaluate the hypothesis concerning carrier exchange between adjacent QDs and to exclude the WL mobility edge or other 3D states one can carry out the experiment where electron-hole (e-h) pairs are photo-injected directly at QDs states. Fig. 2 shows temperature dependent PL quench measured at low excitation condition ($\ll$1 e-h pair/dot), with the photon energy of $\sim$0.918 eV, settled below the WL mobility edge. At high temperatures, exceeding 30 K, PL quenches registered for both structures behave similarly, with one exclusion seen between 30 K and 60 K for InAs/InAlAs QDs that will not be discussed here. One can focus on the low temperature range of 5-30 K, where thermal excitation of e-h pairs to QD excited states or the WL is less efficient, if not negligible. In this range, the PL quench for InAs/InAlAs QDs is not observed at all, however, it is well visible for QDs in the quaternary matrix. This observation can be related to at least two processes: (i) carrier transfer to optically inactive states either located within a neighboring QD or in its vicinity (tail of the WL density of states), (ii) an electron or a hole within an e-h pair is spatially separated in adjacent dots, owing to quantum tunneling effect, without loosing the Coulomb correlation. Both scenarios are very interesting and have not been widely investigated in the InAs/InP material system. One can try to take a look at the second process. An e-h pair, spatially separated between two adjacent dots, should have smaller oscillator strength of the optical transition, translating into long radiative decay time. Therefore, the PL dynamics should be contributed not only by a common e-h recombination but also by the proces postulated above.

Figure 3 shows examples of PL decay traces measured at T=30 K for investigated QD structures at the same excitation conditions. One can see that both profiles are qualitatively different. The PL decay for the InAs/InAlAs QD structure can be well approximated by a mono-exponential process, however, the PL decay profile for InAs/InGaAlAs can be viewed as at least bi-exponential, composed of a fast and a slow component. The fast component of 1.4±0.2 ns decay time is attributed to the e-h recombination at the QD ground state, the

IOP Conf. Series: Journal of Physics: Conf. Series **906** (2017) 012008 doi:10.1088/1742-6596/906/1/012008

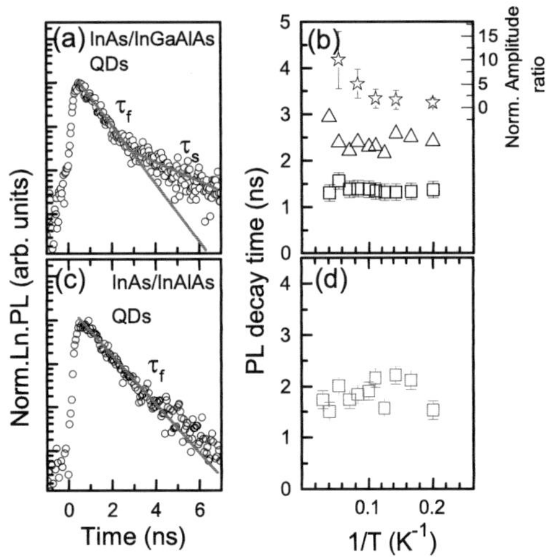

Figure 3. (a),(c) Examples of TRPL traces registered at 0.821 eV and T=30 K for InAs/InGaAlAs QDs and InAs/InAlAs QDs, respectively. (b) The value of PL decay time parameter estimated for fast (τ_f - black squares) and slow (τ_s - black triangles) component of TRPL traces at various temperatures for InAs/InGaAlAs QDs. Black stars indicate the amplitude ratio between fast and slow component. (d) The value of PL decay time parameter for the InAs/InAlAs QDs. Excitation at 0.918 eV.

equivalent process in the InAs/InAlAs QDs has the time constant of 1.9±0.2 ns.[7] Derivation of the slow component by the fitting procedure is quite difficult, nevertheless, the related time constant exceeds 2.4 ns. The slow component can be related to the recombination process of a spatially indirect e-h pair. In the considered temperature range, all components weakly depend on temperature. However, for the InAs/InGaAlAs QDs, the ratio between amplitudes of a slow and a fast component should increase if the carrier exchange process is temperature driven. Indeed, this behavior is observed in the amplitude ratio, as presented in Fig. 3(b).

4. Conclusions

The issue of lateral interdot coupling in a dense ensemble of InAs QDs embedded into different matrix material has been studied by spectroscopic methods at cryogenic temperatures. A striking difference in the spectral and temporal response between $InAs/In_{0.52}Al_{0.48}As$ and $InAs/In_{0.53}Ga_{0.23}Al_{0.24}As$ QDs have been observed and attributed to the presence or absence of an interdot coupling controlled by confinement parameters. Strong temperature driven PL quench in $InAs/In_{0.53}Ga_{0.23}Al_{0.24}As$ QDs and observation of the long-lasting component of the PL decay have suggested the presence of spatially indirect e-h pairs due to coupling between adjacent QDs.

Acknowledgments

This research was supported by The National Science Centre grant MAESTRO No. 2011/02/A/ST3/00152 and Preludium 8 project No. 2014/15/N/ST7/04708. The authors acknowledge technical support from Hamamatsu Co. in near infrared streak camera apparatus.

References

[1] Wang L, Rastelli A, Kiravittaya S, Benyoucef M, Schmidt O G 2009 *Adv. Mater.* **21** 2601-2618
[2] Wu J, Hu X, Lee J, Kim E- S, Wang Z M 2013 *Adv. Optical Mater.* **1** 201-214
[3] Stinaff E A, Scheibner M, Bracker A S, Ponomarev I V, Korenev V L, Ware M E, Doty M F, Reinecke T L, Gammon D 2006 *Science* **311** 636639
[4] Liverini V, Bismuto A, Nevou L, Beck N, and Faist J 2010 *Appl. Phys. Lett.* **97** 221109
[5] Liverini V, Nevou L, Castellano F, Bismuto A, Beck M, Gramm F, and Faist J 2012 *Appl. Phys. Lett.* **101** 261113
[6] Misiewicz J, Sitarek P, Sęk G, and Kudrawiec R 2003 *Mater. Sci.* **21** 263
[7] Dusanowski Ł, Syperek M, Rudno-Rudziński W, Mrowiński P, Sęk G, Misiewicz J, Somers A, Reithmaier J P and Forchel A 2013 *Appl. Phys. Lett.* **103** 253113

Dynamic Rabi oscillations in a quantum dot embedded in a nanobridge in the presence of surface acoustic waves

L Mourokh[1], A Wixforth[2], F Beil[3], M Bichler[4], W Wegscheider[5] and R H Blick[6]

[1]Physics Department, Queens College of CUNY, Queens, NY 11367, USA
[2]Physics Department, University of Augsburg, Universitätsstr. 1, 86159 Augsburg, Germany
[3]Ludwigs-Maximilians-Universität-München, Geschwister-Scholl-Platz 1, 80539 München, Germany
[4]Walter-Schottky-Institute, Am Coloumbwall 3, 85748 Garching, Germany
[5]Laboratory for Solid State Physics, HPF E 7, ETH Zurich, Schafmattstrasse 16, 8093 Zurich, Switzerland
[6]Center for Hybrid Nanostructures (CHyN) and Institutes of Nanostructure and Solid State Physics, University of Hamburg, Lurupuer Chaussee 149, 22761 Hamburg, Germany

lev.murokh@qc.cuny.edu

Abstract. A quantum dot is created within a suspended nanobridge containing a two-dimensional electron gas. The electron current through this dot exhibits well-pronounced Coulomb blockade oscillations. When surface acoustic waves (SAW) are driven through the nanobridge, Coulomb blockade peaks are shifted. To explain this feature, we derive the expressions for the quantum dot level populations and electron currents through these levels and show that SAW-induced Rabi oscillations lead to the observed phenomenology.

1. Introduction

As modern electronic devices become smaller, it is compulsory to investigate possible quantum effects, which emerge due to the electron confinement [1]. In Nano-Electromechanical Systems (NEMS), it is also important to examine how the actual mechanical motion at the nanoscale affects the electrical properties of the structure [2]. The interplay between the electron confinement and nanomechanics is hence of special interest.

Surface acoustic waves (SAWs) are perfectly suited to provide an excitation mechanism for suspended nano-structures [3]. SAWs are acoustic modes propagating at the surface of an elastic solid, having wavelengths in the range of some microns and penetration depths of the same order. Excitation of SAWs at a specific frequency is in particular effective employing a piezoelectric substrate (GaAs in our case) and properly designed interdigitated transducers (IDTs). As typical SAW amplitudes can be in the range of nanometers, they effectively modulate the acoustic properties of nanomechanical devices.

IOP Conf. Series: Journal of Physics: Conf. Series **906** (2017) 012009 doi:10.1088/1742-6596/906/1/012009

2. Sample preparation and experimental results

For sample preparation, we use nanomachined AlGaAs/GaAs-heterostructures including a two-dimensional electron gas, which is grown on a sacrificial layer. Additionally, indentations are used to introduce the lateral confinement, i.e. to form a quantum dot (see Fig. 1). Upon removal of the sacrificial layer, so-called phonon cavities are realized containing the low-dimensional electron system. An additional gate is located above the nanobridge to control the energy levels of the quantum dot. The details of this structure are shown in Fig. 1(a).

We measure the electron current through the system as a function of gate voltage and obtain well-defined Coulomb blockade peaks, as shown Fig. 1(b), confirming the formation of a quantum dot. These measurements are executed at the lattice temperature of 25 mK with an elevated electron temperature of ~100 mK.

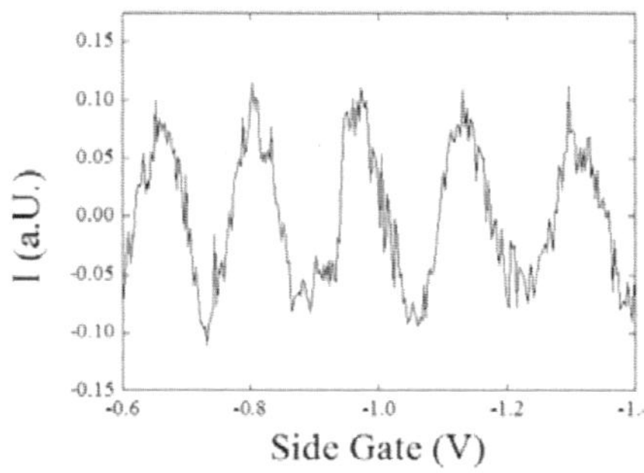

Figure 1. Left: The device under study showing the suspended nanobridge, which contains a 2DEG with the quantum dot defined by the two constrictions left and right. The additional suspended bridge next to the quantum dot is employed as gating electrode for controlling the quantum dot's energy levels. **Right:** Coulomb blockade peaks indicating the single-electron transfer events.

When the beam is subject to SAWs, the positions of the Coulomb blockade peaks are shifted as the power of SAW increases, see Fig. 2.

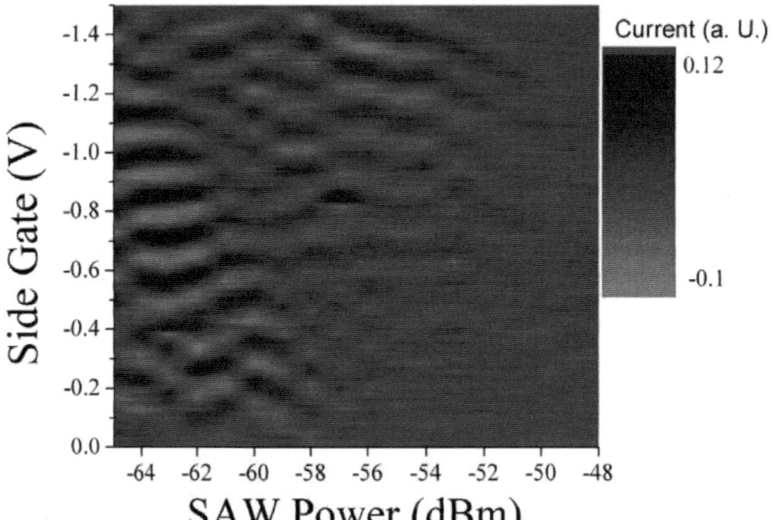

Figure 2. Current through the structure as a function of the side gate voltage and SAW power.

Two key effects were observed. At low SAW power, the positions of the Coulomb blockade peaks are shifted, while at high power the Coulomb blockade is mainly lifted. We attribute the latter effect to the electron heating and will discuss the former one in the next Section.

3. Theoretical approach and discussions

To describe the effect of SAW on electron transport through a quantum dot embedded in a nanobeam, we introduce a term in the Hamiltonian accounting for the motion of the clamping points of the nanobridge. This motion produces a periodic strain ε_{ij} in the beam. Because of a piezoelectric interaction, characterized by the tensor Ξ_{ij}, this strain induces the oscillations of the electron energy in the dot, $\Delta E = -F\sin(\Omega_0 t)$; where Ω_0 is the SAW frequency and F is the amplitude of energy oscillations, $F \sim \Sigma\, \varepsilon_{ij}\, \Xi_{ij}$. Even the moderate SAW amplitude of 0.1nm induces a strain of the order of $\varepsilon \sim 10^{-4}$, which, in turn, results in the amplitude of the energy oscillations of $F \sim 1meV$, if the electron-phonon constant is $\Xi_{ij} \sim 10eV$. It should be noted that our experiment is performed in the adiabatic regime, in contrast to that of Ref. [4], because the frequency of the SAW oscillations, Ω_0, is much smaller than the charging energy of the dot, U, temperature, and the broadening of the electron levels inside the dot, $\Gamma \sim 10\mu eV$. However, the amplitude of the energy oscillations, F, can be comparable to the charging energy.

For simplicity, we examine only one spatial electron wavefunction in the dot, which can be occupied by the two electrons having opposite spin projections, and the corresponding energy levels are separated by the on-site Coulomb energy, U. With spatial level quantization being much smaller than U, other energy levels would be also separated approximately by U, as in well-known Coulomb blockade experiments, and our results can be easily extended. Correspondingly, the Hamiltonian of the system has the form

$$H = \sum_{\sigma=1,2}\left(E_\sigma + F(t)\right)a_\sigma^+ a_\sigma + U a_1^+ a_1 a_2^+ a_2 + \sum_{k,\sigma,\alpha=L,R} E_{k\alpha\sigma} c_{k\alpha\sigma}^+ c_{k\alpha\sigma} - \sum_{k,\sigma,\alpha} T_{k\alpha\sigma} c_{k\alpha\sigma}^+ a_\sigma - \sum_{k,\sigma,\alpha} T_{k\alpha\sigma}^* a_\sigma^+ c_{k\alpha\sigma}. \quad (1)$$

Here, a_σ^+/a_σ and $c_{k\alpha\sigma}^+/c_{k\alpha\sigma}$ are the creation/annihilation operators in the dot and in the α-lead. Using the approach developed in Ref. [5], for the steady state (slowly varying in time with the driving force $F(t)$), we obtain the following equations for the level populations

$$\left(\Gamma_{L\sigma} + \Gamma_{R\sigma}\right)N_{1,2} + \sum_\alpha \Gamma_{\alpha\sigma} N_{2,1}\left[f_{\alpha\sigma}\left(E_\sigma + F(t)\right) + f_{\alpha\sigma}\left(E_\sigma + U + F(t)\right)\right] = \sum_\alpha \Gamma_{\alpha\sigma} f_{\alpha\sigma}\left(E_\sigma + F(t)\right), \quad (2)$$

where $\Gamma_{\alpha\sigma} = 2\pi \sum_k \left|T_{k\alpha\sigma}\right|^2 \delta\left(\omega - E_{k\alpha\sigma}\right)$ are the coupling constants and $f_{\alpha\sigma}$ are the Fermi functions.

Following the same approach, we obtain the expressions for the currents through the σ-level, as

$$I_\sigma = \frac{\Gamma_\sigma}{2}\left[\left(1 - N_{\bar\sigma}\right)\left(f_R\left(E_\sigma + F(t)\right) - f_L\left(E_\sigma + F(t)\right)\right) + N_{\bar\sigma}\left(f_R\left(E_\sigma + U + F(t)\right) - f_L\left(E_\sigma + U + F(t)\right)\right)\right]. \quad (3)$$

The physical meaning of this expression is transparent: if the level $\bar\sigma$ is empty, transport via σ-level occurs at the slowly varying energy $E_\sigma + F(t)$. If the level $\bar\sigma$ is populated, it takes place at the energy $E_\sigma + U + F(t)$. In both cases, the current is proportional to the difference of the Fermi functions of the leads taken at the corresponding energy.

The current averaged over the period of the SAW oscillations, which is measured in the experiment,

$$I_\sigma = \frac{\Omega_0}{2\pi}\int_0^{2\pi/\Omega_0} dt\, I_\sigma(t), \quad (4)$$

is shown in Fig. 3 for the following set of parameters: $U = 0.6$ meV, $\Gamma_{\alpha\sigma} = 0.01$ meV, and the source-drain bias $V_{SD} = 0.001$ meV.

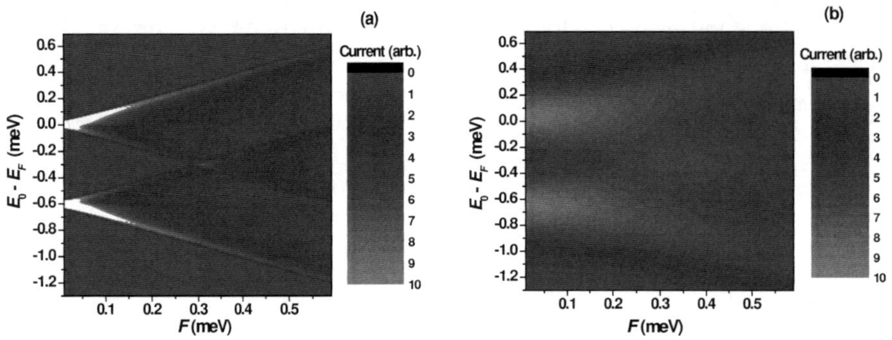

Figure 3. Calculated current through the structure at 100mK (a) and 1.5K (b).

One can see that the Coulomb blockade peaks associated with the two electron levels in the dot become split with the SAW power increased and form a peak in the middle when the lower branch of the upper peak merges with the upper branch of the lower peak resembling the experimental data. This property can be understood from the following point of view. Formally, we have a term $F(t)a_{\sigma}^{+}a_{\sigma}$ in the Hamiltonian, which looks similar to the conventional term describing Rabi oscillations, except it creates and annihilates electrons at the same level. However, the level is broadened by coupling to the leads with $\Gamma >> \Omega_0$, so both levels resonant to the SAW are inside this Γ window. Physically, the electron lifetime in the dot, $1/\Gamma$, is smaller than the period of the SAW oscillations and one electron with the energy $E + \Omega_0$ can emit phonon and escape to the right lead and the second electron with the energy E can come from the left lead and absorb this phonon. Accordingly, we can assume *dynamical* Rabi oscillations in the dot with the phonon-assisted tunneling. This process is the dominating contribution to the averaged current. As a result of the Rabi shift, we have the levels splitting as $E \pm F$ and $E + U \pm F$. At $F = U/2$, two of the levels merge, as can be seen in Fig. 2. It should be emphasized that transport not only probes the level structure, but facilitates the level broadening which is extremely important for this observation of the phonon-assisted tunnelling in the adiabatic regime. It should be noted that the peaks are clearly seen at very low temperature only, being quite broadened already at 1.5 K. It explains the Coulomb blockade lifting at high SAW power seen in the experiment.

4. Summary

In summary, we studied electron transport through a quantum dot formed in the suspended nanobridge, which is mechanically excited via surface acoustic waves. We observe a shift of the Coulomb blockade peaks with increasing SAW power. This can be explained within the framework of "dynamical" Rabi oscillations.

Acknowledgements

The work of LM is partially supported by AFOSR, Award No FA9550-16-1-0279. RHB acknowledges support by the Center for Ultrafast Imaging (CUI) via grant EXC-1074 of the Deutsche Forschungsgemeinschaft (DFG).

References

[1] Ferry D K, Goodnick S and Bird J P 2009 *Transport in Nanostructures* 2nd edition (Cambridge University Press)
[2] Ekinci K L and Roukes M L 2005 *Rev. Sci. Instr.* **76** 061101
[3] Kreft D J and Blick R H 2011 *Surface Acoustic Waves and Nano–Electromechanical Systems, Acoustic Waves - From Microdevices to Helioseismology* ed M G Beghi (InTech)
[4] Naber W J M, Fujisawa T, Liu H W and van der Wiel W G 2006 *Phys. Rev. Lett.* **96** 136807
[5] Mourokh L G, Horing N J M and Smirnov A Yu 2002 *Phys. Rev.* B **66** 085332

IOP Conf. Series: Journal of Physics: Conf. Series **906** (2017) 012010 doi:10.1088/1742-6596/906/1/012010

Hydrodynamic modeling of electron transport in silicon quantum wires

O Muscato[1], **T Castiglione** [1], **A Coco** [2]

[1] Department of Mathematics and Computer Science, University of Catania, Viale A. Doria 6, Catania 95125, Italy

[2] Department of Mechanical Engineering and Mathematical Sciences, Oxford Brookes University, Wheatley campus, OX33 1HX Oxford GB

E-mail: orazio.muscato@unict.it

Abstract. An extended hydrodynamic model self-consistently coupled to the 2D Schrödinger and 3D Poisson equations is introduced, to describe charge transport in Silicon Nanowires. It is been formulated by taking the moments of the multisubband Boltzmann equation, and the closure relations for the fluxes and production terms have been obtained by means of the Maximum Entropy Principle. The low-field mobility for a Gate-All-Around in a SiNW transistor has been evaluated.

1. Introduction

In the last years, the peculiar electro-thermal properties of silicon nanostructures, have attracted an increasing interest because of the potential applications as sensors, field effect transistors, logic gates, thermoelectric generators [1, 2]. Silicon nanowires (SiNWs) are quasi-one-dimensional structures in which electrons are spatially confined in two directions and they are free to move in the orthogonal direction. By shrinking the dimension of these structures, effects of quantum confinement are observed and the wave nature of the electrons must be taken into account. The investigation of these devices by means of numerical simulations can be very informative, provided that a realistic and complete physical model is employed. Under reasonable hypothesis, transport in low-dimension semiconductors can be tackled coupling quantum and semiclassical tools. Therefore, for long channels, semiclassical formulations based on the 1-D Multiband Boltzmann Transport Equation (MBTE) can give reliable simulation results when it is solved self-consistently with the 3-D Poisson and 2-D Schrödinger equations in order to obtain the self-consistent potential and subband energies. We notice that this semiclassical formulation fails if in the channel there are potential barriers: in this case tunneling effects must be taken into account using quantum kinetic or hydrodynamic models [3, 4].

The Monte Carlo method provides a stochastic solution of the MBTE, although affected by statistical noise [3, 5–10]. Another alternative is to take the moments of the MBTE to obtain hydrodynamic-like models, providing a good engineering-oriented approach. Such approach has been successfully used in the electrical and electro-thermal [11–20] simulation of sub-micrometric electronic devices.

2. Quantum confinement and electronic transport

In SiNW the band structure is altered with respect to the bulk case, depending on the cross-section wire dimension, the atomic configuration, and the crystal orientation. Tight-binding (TB) calculations shows that the six equivalent Δ conduction valleys of the bulk Si are split into two groups because of the quantum confinement [21]. The subbands related to the four **unprimed** valleys Δ_4 ([0 ± 10] and [00 ± 1] orthogonal to the wire axis) are projected into a unique valley in the Γ point of the one-dimensional Brillouin zone. Therefore a SiNW is a direct band-gap semiconductor. The subbands related to the **primed** valleys Δ_2 ([± 100] along the wire axis) are found at higher energies and exhibit a minimum, located at $k_x = \pm 0.37\pi/a_0$, and the energy gap between the Δ_4 and Δ_2 bottom valley is 117 meV. The corresponding effective masses m^*, in the parabolic spherical band approximation , are $m^*_{\Delta_2} = 0.94$, $m^*_{\Delta_4} = 0.27$ (in rest electron mass units). For a quantum wire with linear expansion in x-direction, and confined in the plane $y - z$, the normed wave function $\phi(x, y, z)$ can be written in the form

$$\phi(x, y, z) = \chi_l^\mu(y, z) \frac{e^{ik_x x}}{\sqrt{L_x}} \tag{1}$$

where μ is the valley index (one Δ_4 valley and two Δ_2 valleys), $l = 1, N_{sub}$ the subband index, $\chi_l^\mu(y, z)$ is the wave function of the l-th subband and μ-th valley, and the term $e^{ik_x x}/\sqrt{L_x}$ describes an independent plane wave in x-direction confined to the normalization length, $0 \leq x \leq L_x$, with wave number k_x. In the Hartree approximation, a given electron feels the total potential $V_{tot}(x, y, z) = U(y, z) - eV(x, y, z)$, where V is the electrostatic potential, e is the absolute value of the electric charge, and $U(y, z)$ the confining potential. The spatial confinement in the (y, z) plane is governed by the Schrödinger-Poisson system (SP)

$$\begin{cases} H[V]\, \chi_{lx}^\mu[V] = \varepsilon_{lx}^\mu[V]\, \chi_{lx}^\mu[V] \\ H[V] = -\frac{\hbar}{2m_\mu^*}\left(\frac{\partial^2}{\partial y^2} + \frac{\partial^2}{\partial z^2}\right) + V_{tot}(x, y, z) \\ \nabla \cdot [\epsilon_0 \epsilon_r \nabla V(x, y, z)] = -e(N_D - N_A - n[V]) \\ n[V](x, y, z, t) = \sum_\mu \sum_l \rho_l^\mu(x, t) |\chi_{lx}^\mu[V](y, z, t)|^2 \end{cases} \tag{2}$$

where $(2)_1$ is the Schrödinger equation in the Effective Mass Approximation, $(2)_3$ is the Poisson equation, and N_D, N_A are the assigned doping profiles (due to donors and acceptors). The electron density $n[V]$ is given by $(2)_4$, where $\rho_l^\mu(x, t)$ is the linear density in the μ-valley and l-subband which must be evaluated by the transport model (hydrodynamic/kinetic) in the free movement direction. In the following we shall assume a quadratic cross-section of the wire (having total dimension $L_y = L_z$) where the wire is surrounded by an oxide layer which gives rise to a deep potential barrier having $U = 4.05$ eV. The SP system forms a coupled nonlinear PDEs, which it is usually solved by using an adaptive iteration scheme, obtaining the electrostatic potential V, the eigenvalues (or subband energies) ε_{lx}^μ, and the eigenvectors (or electron envelope wavefunctions) χ_{lx}^μ as a function of the unconfined x direction. Then the total electron energy in the μ-valley and l-subband is $E_l^\mu = \varepsilon_{lx}^\mu + \frac{\hbar k_x^2}{2m_\mu^*}$.

The electrons evolve in time under the streaming motion of external forces and spatial gradient, and the randomizing influence of nearly point-like (in space-time) scattering events.

IOP Conf. Series: Journal of Physics: Conf. Series **906** (2017) 012010 doi:10.1088/1742-6596/906/1/012010

For devices with a characteristic length of a few tens of nanometers, the transport of electrons along the axis of the wire can be considered semiclassical within a good approximation [22], and it can be modeled by the 1D MBTE. By taking moments of the MBTE, the following extended hydrodynamic model [23–25] has been obtained

$$\frac{\partial \rho_l^\mu}{\partial t} + \frac{\partial(\rho_l^\mu V_l^\mu)}{\partial x} = \rho_l^\mu \sum_{l'} \mathcal{C}_\rho(\mu, l, \mu, l') + \rho_l^\mu \sum_{l',\mu' \neq \mu} \mathcal{C}_\rho(\mu, l, \mu', l') \tag{3}$$

$$\frac{\partial(\rho_l^\mu V_l^\mu)}{\partial t} + \frac{2}{m_\mu^*}\frac{\partial(\rho_l^\mu W_l^\mu)}{\partial x} + \frac{\rho_l^\mu}{m_\mu^*}\frac{\partial \varepsilon_l^\mu}{\partial x} = \rho_l^\mu \sum_{l'} \mathcal{C}_V(\mu, l, \mu, l') + \rho_l^\mu \sum_{l',\mu' \neq \mu} \mathcal{C}_V(\mu, l, \mu', l') \tag{4}$$

$$\frac{\partial(\rho_l^\mu W_l^\mu)}{\partial t} + \frac{\partial(\rho_l^\mu S_l^\mu)}{\partial x} + \rho_l^\mu \frac{\partial \varepsilon_l^\mu}{\partial x} V_l^\mu = \rho_l^\mu \sum_{l'} \mathcal{C}_W(\mu, l, \mu, l') + \rho_l^\mu \sum_{l',\mu' \neq \mu} \mathcal{C}_W(\mu, l, \mu', l') \tag{5}$$

$$\frac{\partial(\rho_l^\mu S_l^\mu)}{\partial t} + \frac{\partial(\rho_l^\mu F_l^\mu)}{\partial x} + \frac{3\rho_l^\mu}{m_\mu^*}\frac{\partial \varepsilon_l^\mu}{\partial x} W_l^\mu = \rho_l^\mu \sum_{l'} \mathcal{C}_S(\mu, l, \mu, l') + \rho_l^\mu \sum_{l',\mu' \neq \mu} \mathcal{C}_S(\mu, l, \mu', l') \tag{6}$$

where (3)-(6) represent balance equations for the density (ρ_l^μ), momentum (V_l^μ), energy (W_l^μ) and energy-flux (S_l^μ) respectively. This system is not closed due the presence of the high-order moment F_l^μ as well as of the production terms (i.e. the RHS). By using the Maximum Entropy Principle [26–29], one determines functional form of the distribution function in a neighborhood of local thermal equilibrium, and consequently in [24], we have obtained $F_l^\mu = 6(W_l^\mu)^2/m_\mu^*$ and the terms $\mathcal{C}_\rho, \mathcal{C}_V, \mathcal{C}_W, \mathcal{C}_S$, which have been calculated taking into account scattering with acoustic and optical phonons, as well as surface roughness (SR). This model provides a realistic description of the scattering rates, an useful tool to evaluate the low-field mobility and also far-from equilibrium transport regimes.

3. Simulation results

We have considered a Gate-All-Around Silicon Nanowire transistor (SiNWT) having a parallelepiped geometry with a metal gate wrapped around it, in such a way we have a three contact device. Such devices have been designed during these years in order to maintain a good electrostatic control in the channel. The channel has been homogeneously doped to $N_D = 3\ 10^{17}\ \text{cm}^{-3}$ and it is very long (L_x = 120 nm) with respect to the transversal dimensions ($L_y = L_z$ = 12 nm), and the oxide thickness is 1 nm. As preliminary result, we have evaluated the low-field mobility in this long-channel device. In order to evaluate this parameter, we fix an uniform electric field along the channel of 1 kV/cm, and then we compute the mobility as the ratio between the average electron velocity and the driving field, as function of the gate voltage V_g. The average electron velocity has been obtained with the following iterative procedure: i) we fix V_g and solve the Schrödinger-Poisson system (2); ii) once this solution has been obtained, the energies ε_l^μ and wave functions χ_l^μ for each subband are exported into the hydrodynamic model, and the linear density ρ_l^μ is used as initial condition; iii) the hydrodynamic model (3)-(6) is solved and, in the stationary regime, the average velocity has been calculated. In Fig. 1 we plot the electron density in a cross-section of the wire, showing the formation of the surface inversion layer with the electron density peaked close to the oxide interface. In Fig. 2 we show the low-field mobility as function of the gate voltage V_g, obtained by including/excluding the SR mechanism. The inclusion of SR scattering has a strong effect on the low-field mobility dependence, and it is a key point in the simulation of such devices.

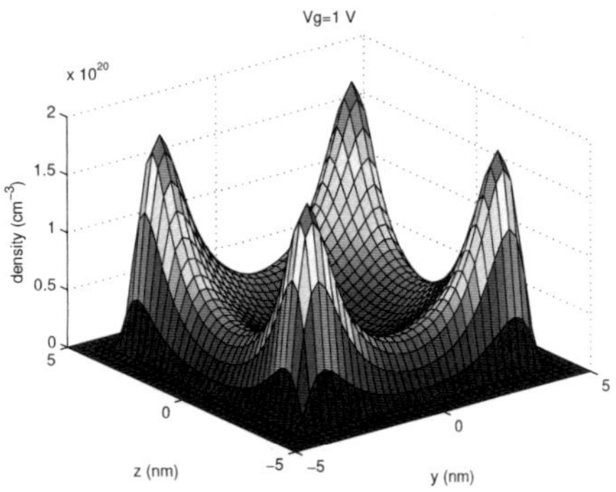

Figure 1. The electron density for $V_g = 1V$ in a wire cross-section.

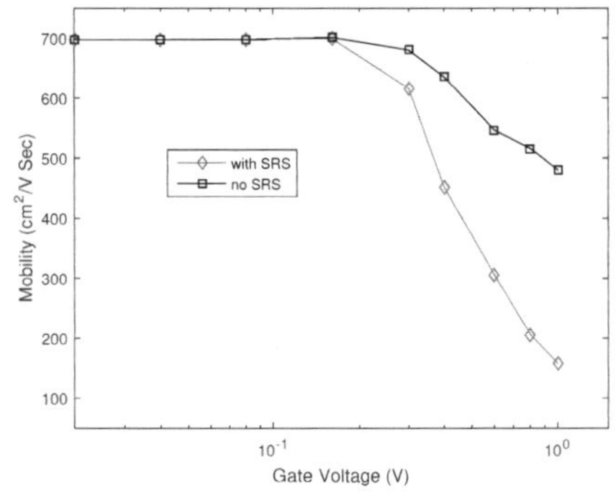

Figure 2. Low-field mobility versus gate voltage, obtained with/without Surface Roughness Scattering mechanism.

Acknowledgments
We acknowledge the support of the Università degli Studi di Catania, FIR 2014 "Charge Transport in Graphene and Low dimensional Structures: modeling and simulation".

References
[1] Pennelli G, Nannini A and Macucci M 2014 *J. Appl. Phys.* **115** 084507
[2] Guerfi Y and Larrieu G 2016 *Nanoscale Res. Lett* **11** 210
[3] Muscato O and Wagner W 2016 *SIAM J. Sci. Comp.* **38** A1483
[4] Gardner C 1994 *SIAM J. Appl. Math.* **54** 409
[5] Lenzi M, Palestri P, Gnani E, Reggiani S, Gnudi A, Esseni D, Selmi L and Baccarani G 2008 *IEEE Trans. elec. Dev.* **55** 2086
[6] Knezevic I, Ramayya E, Vasileska D and Goodnick S 2009 *J. Comput. Theor. Nanosc.* **6** 1725
[7] Ramayya E and Knezevic I 2010 *J. Comput. Electr.* **9** 206
[8] Muscato O, Wagner W and Di Stefano V 2011 *Kinetic Rel. Models* **4** 809
[9] Muscato O, Di Stefano V and Wagner W 2013 *Comput. Math. Appl.* **65** 520
[10] Ryu H 2016 *Nanos. Res.Lett.* **11** 36
[11] Muscato O and Di Stefano V 2011 *J. Phys. A:Math. Theor.* **44** 105501
[12] Muscato O and Di Stefano V 2011 *J. Stat. Phys.* **144** 171
[13] Muscato O and Di Stefano V 2011 *J. Appl. Phys.* **110** 093706
[14] Di Stefano V and Muscato O 2012 *Acta Appl. Math.* **121** 225
[15] Muscato O and Di Stefano V 2013 *Semicond. Sci. Technol.* **28** 025021
[16] Muscato O, Wagner W and Di Stefano V 2014 *COMPEL* **33** 1198
[17] Mascali G 2015 *Eur. J. Appl. Math.* **26** 477
[18] Muscato O and Di Stefano V 2015 *SIAM J. Appl. Math.* **75** 1941
[19] Mascali G 2016 *J. Stat. Phys.* **163** 1268
[20] Mascali G 2017 *J. Comput. Electr.* **16** 180
[21] Zheng Y, Rivas C, Lake R, Alam K, Boykin T and Klimeck G 2005 *IEEE Trans. Elec. Dev.* **52** 1097
[22] Ferry D, Goodnick S and Bird J 2009 *Transport in Nanostructures* (Cambridge: Cambridge Univ. Press)
[23] Muscato O and Castiglione T 2016 *Comm. Appl. Industr. Math.* **7** 8
[24] Muscato O and Castiglione T 2016 *Entropy* **18** 368
[25] Castiglione T and Muscato O 2017 *J. Comput. Theor. Transp.* **46** 186
[26] Camiola V, Mascali G and Romano V 2012 *Contin. Mech. Thermodyn.* **24** 417
[27] Muscato O and Di Stefano V 2012 *J. Comp. Electr.* **11** 45
[28] Camiola V, Mascali G and Romano V 2013 *Math. Comput. Mod.* **58** 321
[29] Muscato O and Di Stefano V 2014 *Contin. Mech. Thermodyn.* **26** 197

IOP Conf. Series: Journal of Physics: Conf. Series **906** (2017) 012011 doi:10.1088/1742-6596/906/1/012011

Efficient Monte Carlo-based algorithms for the Wigner transport equation

Orazio Muscato and Vincenza Di Stefano

Department of Mathematics and Computer Science, University of Catania, Viale A. Doria 6, Catania 95125, Italy

E-mail: orazio.muscato@unict.it

Abstract. The Wigner transport equation is solved by Direct Simulation Monte Carlo, based on the generation and annihilation of signed particles. In this framework, stochastic algorithms are derived using the theory of pure jump processes with a general state space. Numerical experiments on benchmark test cases are shown.

1. Introduction

The continuous scaling down of semiconductor devices is nowadays at a point, where active lengths are of the order of only a few tens of nanometers. Effects such as particle tunneling through source-to-drain potential profiles are now highly relevant and cannot be ignored. From this perspective, only full quantum models are capable of describing the appropriate physics. A well-known model is the Wigner equation which can be augmented by a Boltzmann-like collision operator accounting for the process of decoherence. The Wigner equation writes [1]:

$$\frac{\partial f_W}{\partial t} + \frac{\hbar}{m^*} k \cdot \nabla_x f_W = \mathcal{Q}(f_W) \quad , \quad \mathcal{Q}(f_W) = \int V_W(x, k-k') f_W(x,k')\, dk' \tag{1}$$

$$V_W(x,k) = \frac{1}{i\hbar(2\pi)^d} \int dx'\, e^{-ik\cdot x'} \left[V\left(x + \frac{x'}{2}\right) - V\left(x - \frac{x'}{2}\right)\right] \tag{2}$$

where V_W is the Wigner potential, and $V(x)$ is the potential energy. The Wigner potential is a non-local potential operator which is responsible of the quantum transport, it is real-valued, and anti-symmetric with respect to k.

The numerical solution of the Wigner equation can be obtained using high-order finite difference solvers [2–5]. To avoid discretization problems, particle Monte Carlo (MC) methods can be used (see [6] for a review), despite the large computational times.

In the realm of the MC methods, the *Signed particle Monte Carlo approach* [7, 8] seems to be the most promising, because it can be understood in a probabilistic framework. In fact, the quantum evolution term $(1)_2$ is interpreted as a *Gain* term of a collisional operator in which the *Loss* term is missing. But the Wigner potential (2) is not always positive and cannot be considered a scattering term. For this reason, it can be separated into a positive and negative parts $V_W^+, V_W^- \geq 0$ such that $V_W = V_W^+ - V_W^-$. Therefore the quantum evolution term can be written as

$$\mathcal{Q}(f_W) = \int dk' w(k',k) f_W(x,k') - \gamma(x) f_W(x,k) \quad , \quad \gamma(x) = \int dk' \, V_W^+(x, k-k') \tag{3}$$

$$w(k',k) = V_W^+(x,k-k') - V_W^-(x,k-k') + \gamma(x)\delta(k-k') \tag{4}$$

where now $\mathcal{Q}(f_W)$ is written as the difference of *Gain* and *Loss* terms. The corresponding scattering mechanism is dictated by (4), where the first term represents the creation of a new positive particle, the second the creation of a negative one and the last term means that the initial particle is maintained. Hence a real-valued weight (or affinity) must be introduced, to describe this pair production. The intensity of the pair creation is ruled by γ, and the momentum of the new particles generated according to the probability V_W^+. The main drawbacks of this procedure reside in the exponential grow of the particle number, and in the evaluation of the γ and V_W^+ functions because they are rapidly oscillating.

This creation process has been recently understood in terms of the "piecewise deterministic Markov processes" theory (PDMP) [9, 10]. Each particle is characterized by a real-valued weight A, a position x and a wave-vector k. The particle position changes continuously, according to the velocity determined by the wave-vector. New particles (called offspring) are created randomly and added to the system. The main result is that appropriate functionals of this stochastic process satisfy a weak form of the Wigner equation. Moreover, this approach has certain advantages compared to other derivations. In particular, it suggests a variety of new algorithms as well as some of the algorithms previously considered in the literature.

2. New generation algorithms

According to PDMP, we shall introduce a new algorithm, which based on a majorant of the Wigner potential

$$|V_W(x,k)| \le \hat{V}_W(x,k) \quad \forall x,k \quad , \quad \hat{\gamma}(x,c) = \frac{1}{2}\int_{\mathbb{B}} \hat{V}_W(x,k)\, dk \tag{5}$$

where $\mathbb{B} = \left\{ k \in \mathbb{R}^d : \|k\| \le c \right\}$ and c is a cutoff. A splitting time step $\Delta t \le 1/\hat{\gamma}$ is used in order to separate the transport and the creation processes. The new algorithm performs the evolution of the system on the time interval $[0, \Delta t]$ according to the following steps:

1. Transport step : the particle positions change according to

$$x_j := x_j + \frac{\hbar}{m^*} k_j \, \Delta t \quad , \quad j = 1, ..., N$$

2. Creation step : for $j = 1, .., N$ new particles are created according to the following rules:
   (a) with probability $1 - \hat{\gamma}(x_j, c)\Delta t$ do not create anything.
   (b) Otherwise generate randomly a new $\tilde{k}$ according to the probability

$$\frac{1}{2\hat{\gamma}(x_j, c)} \hat{V}_W(x_j, k), \quad k \in \mathbb{B}$$

   (c) with probability

$$1 - \frac{|V_W(x_j, \tilde{k})|}{\hat{V}_W(x_j, \tilde{k})}$$

   do not create anything.

(d) Otherwise create the particle couple

$$[\tilde{A}(A_j, x_j, \tilde{k}_j), x_j, k_j + \tilde{k}] \quad , \quad [-\tilde{A}(A_j, x_j, \tilde{k}_j), x_j, k_j - \tilde{k}]$$

with $\tilde{A}(A, x, k) = A \,\mathrm{sign}\, V_W(x, k)$. Put $N := N + 2$.

3. Cancellation step: if $N \geq N_{canc}$ pairs of particles with opposite affinity, which belong to the same element of the phase-space are removed.

The particle creation rate, in this new algorithm, is dictated by the function $\hat{\gamma}$ $(5)_2$ obtaining a gain factor 2 with respect to standard creation procedure employing the function γ $(3)_2$; moreover, the new algorithm has the advantage that the calculation of the rate function γ, which is highly oscillating, is avoided. Recently we have developed another MC algorithm (which we shall call no-splitting), performing the particle evolution on a finite time interval $[0, T]$ without any time discretization error, i.e. avoiding the time step splitting error [11].

3. Benchmark studies

As benchmark, we have considered a potential barrier with the shape of a gaussian function, centered at x=0 with dispersion $\sigma = 1$ nm and height $a = 0.3$ eV. The Wigner potential (2), in the 1D case, writes

$$V_W(x, k) = \frac{2a\sigma\sqrt{2\pi}}{\pi^- h} \exp\left[-2(\sigma k)^2\right] \sin(2kx) \quad . \tag{6}$$

The creation algorithm is based on the following majorant of $V_W(x, k)$

$$\hat{V}_W(x, k) = \frac{2a\sigma\sqrt{2\pi}}{\pi^- h} \quad , \quad \hat{\gamma}(x, c) = \frac{2a\sigma\sqrt{2\pi}}{\pi^- h} c \tag{7}$$

The initial condition for the Wigner equation is

$$f_W(x, k, 0) = \frac{1}{\pi} \exp\left[-\frac{(x - x_0)^2}{2\sigma_0^2}\right] \exp\left[-2\sigma_0^2(k - k_0)^2\right] \tag{8}$$

where x_0 = -15 nm, k_0= 0.7 nm^{-1}, σ_0 = 2.852 nm, and we have chosen absorption boundary conditions. In the x-space we have considered an uniform mesh [-30, 30](nm) with N_x = 200 grid-points; also in the k-space we have an uniform mesh [-10, 10](nm^{-1}) with N_k=400. The cutoff has been fixed c = 8 nm^{-1}, the initial particle number is N_{ini} = 160000, the cancellation parameter N_{canc} = 480000. Since for pure states the Wigner and Schrödinger equations are completely equivalent, we have compared the MC mean density with that obtained using a high-order deterministic Schrödinger equation solver [12]. The mean density, obtained at 20 fsec. with the two solvers, is shown in figure 1, for various time steps. We can observe that the MC solution converge to the Schrödinger one, showing clearly the error introduced by the time step. In figure 2 we plot the mean density obtained using the no-splitting algo (without any time step) and the Schrödinger one: the agreement between the two mean densities is very good. Figure 3 shows the CPU times versus the time step as well as the CPU obtained with the no-splitting algo, and, in figure 4, the max absolute error between the MC and the Schrödinger solutions. From these figures it is clear the advantage to use the no-splitting algorithm instead of standard ones.

4. Conclusions

The Wigner equation has been solved by using the Signed particle Monte Carlo method where, new pair of particles characterized by a sign, are created randomly and added to the system. This creation mechanism has been understood in terms of the Markov jump process. Numerical experiments have been performed successfully on a benchmark test case. Future researches will include simulation of silicon nanowires according to the guidelines in [13–15].

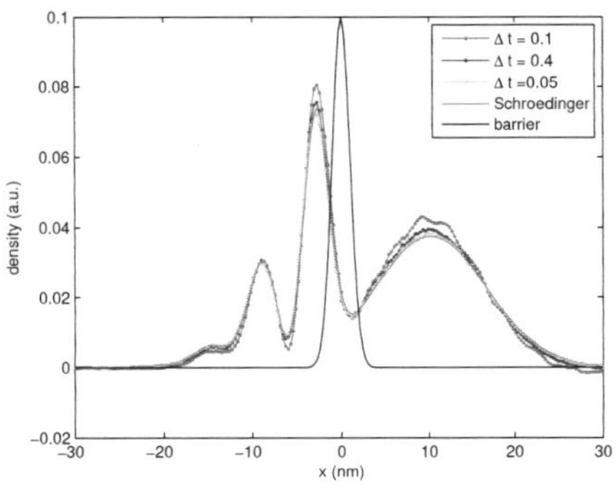

Figure 1. The mean density versus position at simulation time 20 fsec., obtained with some time steps Δt.

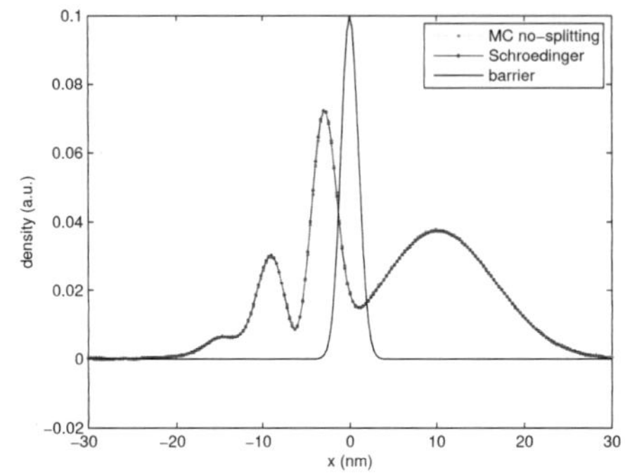

Figure 2. The mean density versus position at simulation time 20 fsec., obtained with the no-splitting algorithm.

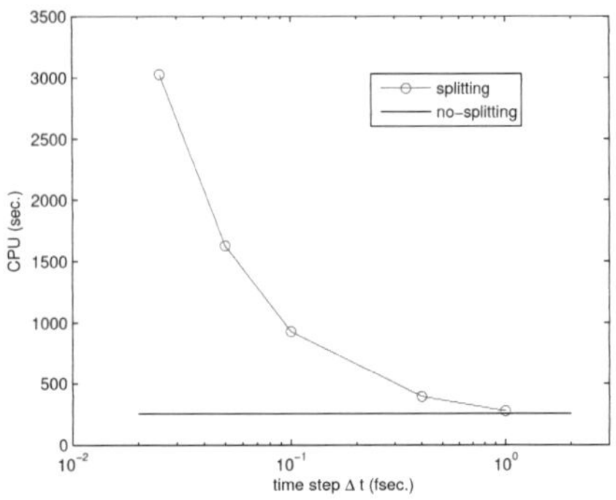

Figure 3. CPU time versus the time step Δt.

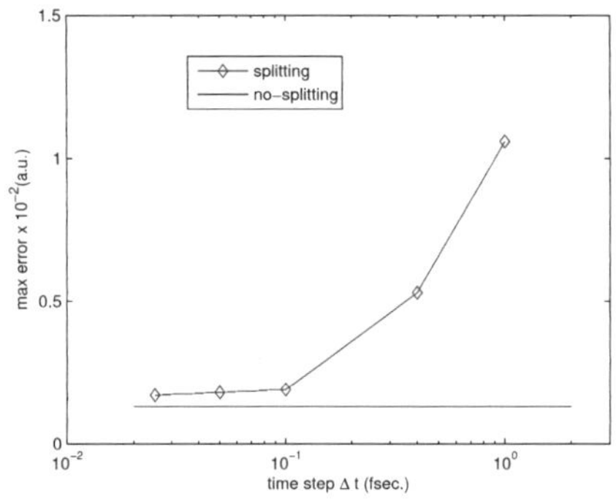

Figure 4. Max error versus the time step Δt.

Acknowledgments

We acknowledge the support of the Università degli Studi di Catania, FIR 2014 "Charge Transport in Graphene and Low dimensional Structures: modeling and simulation".

References

[1] Querlioz D and Dollfus P 2010 *The Wigner Monte Carlo method for nanoelectronics devices* (Wiley)
[2] Morandi O and Demeio L 2008 *Transp. Theor. Stat. Phys.* **37** 473
[3] Morandi O and Schuerrer F 2011 *J. Phys. A: Math. Theor.* **26** 265301
[4] Shao S, Lu T and Cai W 2011 *Comm. Comput. Phys.* **9** 711
[5] Dorda A and Schuerrer F 2015 *J. Comp. Electr.* **284** 95
[6] Kosina H 2006 *Comp. Scien. Engineer.* **2** 100
[7] Nedjalkov M, Kosina H, Selberherr S, Ringhofer C and Ferry D 2004 *Phys. Rev. B* **70** 115319
[8] Shao S and Sellier J 2015 *J. Comp. Phys.* **300** 167
[9] Wagner W 2016 *Kin. Rel. Models* **9** 217
[10] Muscato O and Wagner W 2016 *SIAM J. Sci. Comput.* **38** A1483
[11] Muscato O and Wagner W 2017 *A stochastic algorithm without time discretization error for the Wigner equation* **Preprint**
[12] Chin S and Chen C 2001 *J. Chem. Phys.* **114** 7338
[13] Muscato O and Castiglione T 2016 *Comm. Appl. Industr. Math.* **7** 8
[14] Muscato O and Castiglione T 2016 *Entropy* **18** 368
[15] Castiglione T and Muscato O 2017 *J. Comput. Theor. Transp.* **46** 186

IOP Conf. Series: Journal of Physics: Conf. Series **906** (2017) 012012 doi:10.1088/1742-6596/906/1/012012

Ultrafast Control of Electron Dynamics in 3D Topological Insulator

S. Azar Oliaei Motlagh, Jhih-Sheng Wu, Vadym Apalkov, and Mark I. Stockman

Center for Nano-Optics (CeNO) and Department of Physics and Astronomy, Georgia State University, Atlanta, Georgia 30303, USA
oliaei@mail.phy-astr.gsu.edu

Abstract. We theoretically study the coherent dynamics of Dirac fermions at the surface of a 3D topological insulator (Bi_2Se_3) in the field of an ultrafast laser pulse. In the presence of the ultrafast pulse, which has a femtosecond time scale that is less than the characteristic electron scattering time in these materials, the electron dynamics is coherent. Because of the gapless dispersion relation, the electrons dynamics is highly irreversible, which is dramatically different from dielectrics (fused silica, quartz, and sapphire). Due to irreversibility, finite conduction band population does exist after the pulse ends. Within two-band k.p Hamiltonian, which includes cubic momentum hexagonal warping terms, the residual conduction band (RCB) population in the reciprocal space is highly anisotropic. The distribution of RCB population in the reciprocal space strongly depends on the polarization of the ultrafast laser pulse.

1. Introduction

With the invention of the laser pulse and the progress in nonlinear optics, it becomes possible to use a few cycle strong optical pulses to study the effects of high fields (several volts per angstrom) on electron dynamics in different materials [1,2]. Moreover, this progress makes it possible to probe a broad range of electronic dynamics on the atomic scale with the accuracy of the order of 100 as or even better [1,2].

In this paper, we study ultrafast electron dynamic on the surface states of a 3D topological insulator (Bi_2Se_3). Topological insulators (TIs) are a new class of materials which have the bulk band gap like ordinary insulators. This bulk band gap is due to a strong spin-orbit coupling in these materials [3,4]. However, unlike the ordinary insulators, TIs have conducting edge states or conducting surface states, depending on the dimensionality of the materials [3,4]. These metallic surface states (or edge states) for a 3D TI (or a 2D TI) are protected by time-reversal symmetry [3].

We consider an ultrafast pulse normally incident on the surface of a 3D topological insulator. Under the field of the pulse, which has duration of about 5 fs, the electron dynamics is coherent. In this work, we show that the electron dynamics is irreversible and anisotropic, which is due to cubic momentum hexagonal warping terms in the Hamiltonian.

2. Theory and model description

We describe the surface states of a 3D TI (Bi_2Se_3) near the Γ point within the following effective Hamiltonian, which was derived from the k.p perturbation theory [3]:

$$H_0 = A_1 + A_2 k_{\|}^2 + A_3(\sigma_x k_y - \sigma_y k_x) + A_4(k_+^3 + k_-^3)\sigma_z \tag{1}$$

where $k_\pm = k_x \pm i k_y = k_{\|} e^{\pm i\theta}$, σ_x, σ_y, and σ_z are Pauli matrices, and constants A_1, A_2, A_3, and A_4 are equal to 0.034 (eV), 23.725 (eV. Å^2), 3.297 (eV.Å), and 25.045 (eV. Å^3), respectively. Within the effective Hamiltonian (1), we study the interaction of an electron system with a normally incident ultrashort laser pulse, which has the following profile:

$$F(t) = F_0 e^{-u^2}(1-2u^2) \tag{2}$$

where $F(t)$ is the electric field, F_0 is the amplitude of the pulse, $u = t/\tau$, and τ=1 fs is the pulse duration, which corresponds to the carrier frequency $\hbar\omega = 1.5\,\text{eV}$ [5].

In the presence of the pulse's field, the time-dependent Hamiltonian of an electron is written in the following form:

$$H(t) = H_0 - e\boldsymbol{F}(t).\boldsymbol{r}$$

where $\boldsymbol{F}(t) = (F_x(t), F_y(t))$, $\boldsymbol{r} = (\text{x}, \text{y})$, and e is the electron charge. The time-dependent Schrodinger equation, $i\hbar \frac{d\psi}{dt} = H(t)\psi$, describes nonlinear electron dynamic. The general solution of this equation is expressed in the basis of the Houston functions ($\Phi_{\alpha\boldsymbol{q}}^{(H)}(\boldsymbol{r},t)$) as

$$\psi_{\boldsymbol{q}}(\boldsymbol{r},t) = \sum_{\alpha=c,v} \beta_{\alpha\boldsymbol{q}}(t)\Phi_{\alpha\boldsymbol{q}}^{(H)}(\boldsymbol{r},t) \tag{3}$$

Where $\Phi_{\alpha\boldsymbol{q}}^{(H)}(\boldsymbol{r},t) = \psi_{\boldsymbol{k}(\boldsymbol{q},t)}^{\alpha}(\boldsymbol{r}) e^{-i/\hbar \int_{-\infty}^{t} E[\boldsymbol{k}(\boldsymbol{q},t)]dt_1}$ is the solution for the Schrodinger equation for a single band α, i.e., without the interband coupling. Here $\beta_{\alpha\boldsymbol{q}}(t)$ $(\alpha = c, v)$ are expansion coefficients, which satisfy the following system of coupled ordinary differential equations:

$$\begin{cases} \dfrac{d\beta_{c\boldsymbol{q}}(t)}{dt} = -i\dfrac{\boldsymbol{F}(t).\boldsymbol{Q}_{\boldsymbol{q}}(t)}{\hbar}\beta_{v\boldsymbol{q}}(t) \\ \dfrac{d\beta_{v\boldsymbol{q}}(t)}{dt} = -i\dfrac{\boldsymbol{F}(t).\boldsymbol{Q}_{\boldsymbol{q}}^{*}(t)}{\hbar}\beta_{c\boldsymbol{q}}(t) \end{cases} \tag{4}$$

Where $\boldsymbol{Q}_{\boldsymbol{q}}(t) = \boldsymbol{D}[\boldsymbol{k}(\boldsymbol{q},t)] e^{-\frac{i}{\hbar}\int dt_1 \{E_c[\boldsymbol{k}(\boldsymbol{q},t_1)] - E_v[\boldsymbol{k}(\boldsymbol{q},t_1)]\}}$ and $\boldsymbol{D}(\boldsymbol{k}) = \left\langle \psi_{\boldsymbol{k}}^{c} \middle| e\boldsymbol{r} \middle| \psi_{\boldsymbol{k}}^{v} \right\rangle$ is the interband dipole matrix element. The interband dipole matrix elements have the following form:

$$D_x = N\left(-\frac{1}{2}\frac{k_y}{k_x^2 + k_y^2} - i\frac{2k_x^4 + 3k_y^2(k_x^2 - k_y^2)}{(k_x^2 + k_y^2)\sqrt{4k_x^2(k_x^2 - 3k_y^2)^2 + (A_3/A_4)^2(k_x^2 + k_y^2)}} \right) \tag{5}$$

$$D_y = N\left(\frac{1}{2}\frac{k_x}{k_x^2 + k_y^2} + i\frac{k_x k_y(7k_x^2 + 3k_y^2)}{(k_x^2 + k_y^2)\sqrt{4k_x^2(k_x^2 - 3k_y^2)^2 + (A_3/A_4)^2(k_x^2 + k_y^2)}} \right) \tag{6}$$

Where $N = e\left(1 + \dfrac{4k_x^2(k_x^2 - 3k_y^2)^2}{(A_3/A_4)^2(k_x^2 + k_y^2)}\right)^{-1/2}$.

Within a single band, the Bloch acceleration theorem governs the intraband electron dynamics:

$$\hbar\frac{d\boldsymbol{k}}{dt} = e\boldsymbol{F}(t) \tag{7}$$

With the initial wave vector **q**, the time-dependent wave vector, which is the solution of Eq. (7), is given by the following expression:

$$\boldsymbol{k}(\boldsymbol{q},t)=\boldsymbol{q}+\frac{e}{\hbar}\boldsymbol{A}(t), \qquad \boldsymbol{A}(\mathrm{t})=\int_{-\infty}^{t}\boldsymbol{F}(t_1)dt_1. \tag{8}$$

We solved the system of coupled equations (4) with the assumption that the valence band is initially occupied and the conduction band is empty. The electron dynamics is irreversible, which is characterized by the residual conduction band (RCB) population ($RCB=\left|\beta_{cq}\right|^2$), i.e., conduction band population after the pulse ends.
Electric current, J, is generated during the pulse and is defined by the following expression:

$$J(t)=J_x(t)\cos(\theta)+J_y(t)\sin(\theta), \tag{9}$$

where $J_p(t)=\frac{e}{a^2}\sum_q\sum_{\alpha_1,\alpha_2=v,c}\beta^*_{\alpha_1 q}V_p^{\alpha_1\alpha_2}\beta_{\alpha_2 q}(t)$,

p = x,y, $V_p^{\alpha_1\alpha_2}$ are the matrix elements of the velocity operators $\hat{V}_p=\frac{1}{\hbar}\frac{\partial H_0}{\partial k_p}$.

3. Results and discussion

For linearly polarized pulse, the distribution of the residual conduction band population in the reciprocal space is shown in Fig.1 for different polarizations (ϴ) of the pulse. These distributions show that RCB populations have high contrast peaks near the local maxima of the dipole matrix element. Moreover, RCB population is highly anisotropic. The conduction band population distribution is characterized by hot spots, the number of which depends on polarization of the pulse.

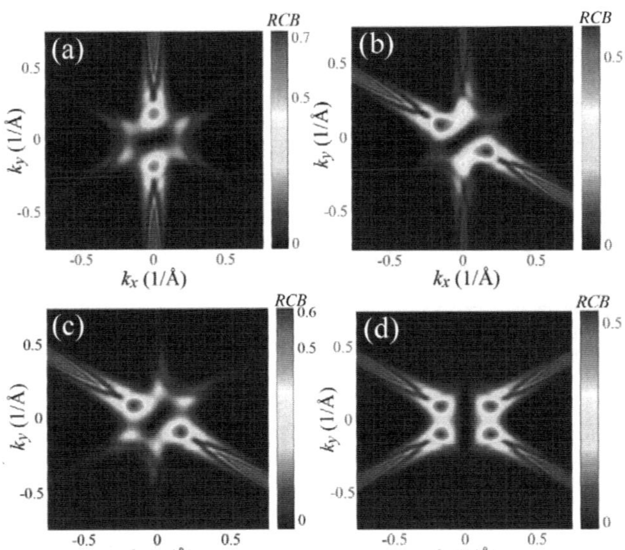

FIG. 1. Distribution of the residual conduction band population in reciprocal space for linearly polarized pulse with the amplitude of 0.1 (V/Å) (a) ϴ = 0; (b) ϴ = π/4; (c) ϴ = π/3; (d) ϴ = π/2;

In Fig. 2(a), the total conduction band population is shown as a function of time for different polarizations of the optical pulse. This figure indicates that the electron dynamics is highly irreversible and is characterized by large RCB population, which is comparable to the maximum conduction band population during the pulse.

During the pulse, the asymmetric distribution of finite conduction band population generates the electric current which can be found from Eq. (9). In Fig. 2 (b), the electric current is shown as a function of time for different polarizations of the optical pulse. Here the pulse amplitude is 0.1 (V/Å). The vector potential, which is the integral of the electric field is also shown in the figure 2 (b).

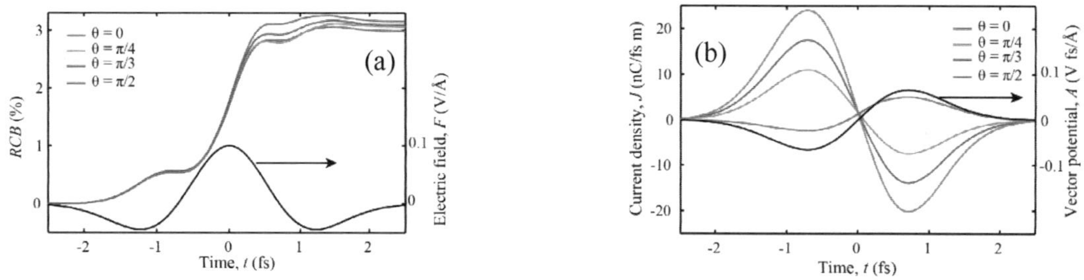

FIG. 2. (a) Total conduction band population as a function of time. The time dependent electric field of the pulse is shown by black line. (b) Electric current as a function of time and time dependent vector potential. The results are shown for different polarizations of the optical pulse: different polarizations Θ = 0, Θ = π/4, Θ = π/3, Θ = π/2. The amplitude of the pulse is 0.1 V/Å.

4. Conclusion

We study ultrafast electron dynamics of a 2D electron system on the surface of a 3D topological insulator. The electron dynamics in such system is highly irreversible and anisotropic. The irreversibility of the electron dynamics is due to zero band gap of the surface states. In this case, the interaction of the laser pulse with the system satisfies the nonadiabatic condition, $\omega_0 \gg \Delta_g/\hbar$, where ω_0 is the laser frequency and Δ_g is the band gap. The anisotropy is due to cubic terms in the effective Hamiltonian of the system, which results in the anisotropy of the interband dipole matrix elements near the Dirac point. Such anisotropy produces hot spots in the distribution of the conduction band population in the reciprocal space, where the number of the spots depends on the polarization of the optical pulse.

Redistribution of the electrons between the conduction and valence bands generates the electric current during the pulse. The magnitude of the current is also anisotropic and depends on the polarization of the pulse.

Acknowledgment
Major funding was provided by Grant No. DE-FG02-01ER15213 from the Chemical Sciences, Biosciences and Geosciences Division, Office of Basic Energy Sciences, Office of Science, US Department of Energy. Numerical simulations have been performed using support by Grant No. DE-FG02-11ER46789 from the Materials Sciences and Engineering Division of the Office of the Basic Energy Sciences, Office of Science, U.S. Department of Energy. The work of V.A. was supported by Grant No. ECCS-1308473 from NSF.

References

1. Weiner, A. M. Ultrafast optical pulse shaping: A tutorial review. *Opt. Commun.* **284,** 3669–3692 (2011).
2. Schultze, M. *et al.* Controlling dielectrics with the electric field of light. *Nature* **493,** 75–78 (2012).
3. Liu, C. *et al.* Model Hamiltonian for Topological Insulators. 32–34 (2011).
4. Fu, L. & Kane, C. L. Topological insulators with inversion symmetry. *Phys. Rev. B* **76,** 045302 (2007).
5. Oliaei Motlagh, S. A., Apalkov, V. & Stockman, M. I. Interaction of crystalline topological insulator with an ultrashort laser pulse. *Phys. Rev. B* **95,** 085438 (2017).

Monte Carlo Simulation of THz Radiation Detection in GaN MOSFET n^+nn^+ Channel with Uncentered Gate in n-region

C. Palermo[1], J. Torres[1], L. Varani[1], V. Gružinskis[2], E. Starikov[2], P. Shiktorov[2], S. Ašmontas[2],and A. Sužiedelis[2]

[1]Institute of Electronics and Systems, University of Montpellier, Montpellier 34095, France

[2]Electron. Dep., National Center for Phys. Sci. and Technol. , Vilnius LT 10257, Lithuania

*Email: luca.varani@umontpellier.fr

Abstract. Electron transport and drain current noise in the wurtzite GaN MOSFET have been studied by Monte Carlo particle simulation which simultaneously solves the Boltzmann transport and pseudo-2D Poisson equations. A proper design of GaN MOSFET n^+nn^+ channel with uncentered gate in n-region to reach the maximum detection sensitivity is proposed. It is shown that the main role in formation of longitudinal transport asymmetry and THz radiation detection is played by optical phonon emission process. It is found that the detection current at 300 K is maximal in frequency range from 0.5 to 7 THz. At higher frequenciea the detection current rapidly decreases due to the inertia of electron motion.

1. Introduction

To detect the microwave power semiconductor diodes with asymmetric current-voltage relation (CVR) are widely used. Usually, such an asymmetry is obtained by forming various barriers (pin, Schottky, heterojunction, etc.) in the active region. Unfortunately, in going to terahertz (THz) region the sensitivity of barrier-structures rapidly decreases due to time-delay related with large values of the barrier capacitance and its resistance. In contrast, as it was experimentally demonstrated [1], these restrictions can be overcame in detector based on asymmetricaly necked nn+ GaAs planar structure. The detector operated in wide frequency band (30 GHz - 2.5 THz) at room temperature [1]. Also the detector based on n+nn+ GaAs diode with nonuniform doped n region for room temperature operation in 20 GHz - 3 THz frequency band is proposed and simulated by Monte Carlo particle (MCP) technique [2]. It is shown in ref.2 that upper limit of detection range depends on optical phonon emission rate. The drain current vs drain voltage relation assymmetry needed for the microwave power detection in field effect transistors can be easily controlled by the gate position on the channel. The optical phonon emission rate in GaN considerably exceeds the one in GaAs.

The aim of this article is to study the possibility of THz detection in wurtzite GaN MOSFETs at room temperature. The electron transport and noise is simulated by the Monte Carlo particle (MCP) technique coupled to a pseudo-2D Poisson solver [3]. The main attention is paid to the optimum conditions for detection in the THz frequency range by varying the MOSFET geometry and doping. The mechanism responsible for THz detection is discussed from a microscopic point of view.

IOP Conf. Series: Journal of Physics: Conf. Series **906** (2017) 012013 doi:10.1088/1742-6596/906/1/012013

2. Model

Below we present the calculations of electron transport and noise in optimized for THz detection MOSFET structure based on a 50–450–50 nm n^+nn^+ wurtzite GaN 50 nm wide channel. A 200 nm gate is placed in the n-region 200 nm away from n^+ source and 50 nm from the n^+ drain at a distance of 25 nm from the channel. The doping concentrations are $n = 10^{17}$ cm^{-3} and $n^+ = 2\times10^{18}$ cm^{-3}. For Monte Carlo simulation the wurtzite GaN band and material parameters of a spherically symmetric nonparabolic conduction band are taken from Ref. [4]. Electron scatterings by polar optical and acoustic deformation phonons, and ionized impurities are included. The common source configuration is considered. All the calculations are made for 300 K temperature an gate bias $U_g = 0$. V .The number of simulated particles, depending on the case, is varied from 10^5 to 10^6. The time step in all cases was 0.5 fs.

3. Steady state

The drain current - drain voltage (I-V) relation calculated in GaN MOSFET is presented in Fig. 1. For better representation of the I-V relation asymmetry the negative values of bias and current are changed to positive ones. The asymmetry of I-V relation is evident from Fig. 1 where the absolute value of drain current at negative drain bias exceed the one at positive bias. In Fig. 2 the doping and electron concentration profiles in GaN MOSFET channel at U_d = - 0.2 and 0.2 V are shown. One can see in

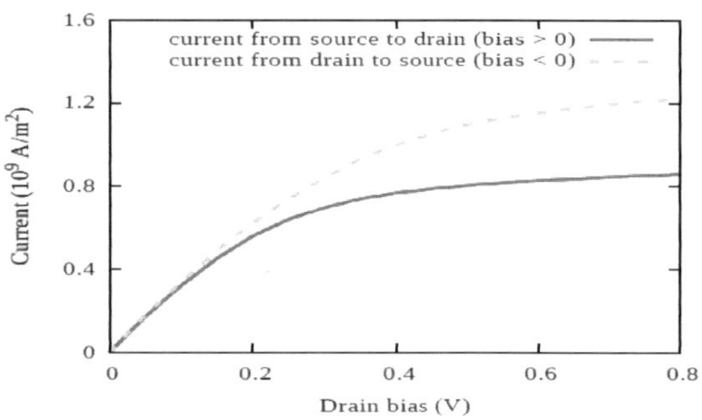

Figure 1. GaN MOSFET I-V relation at drain bias from -0.8 to 0.8 V.

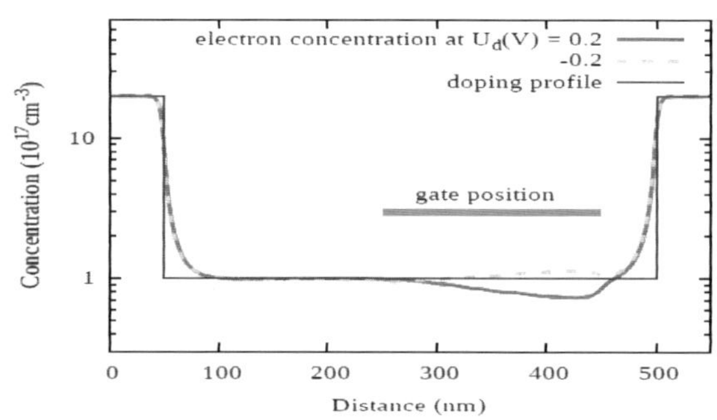

Figure 2. The doping and electron concentration profiles in MOSFET channel at U_d = – 0.2 and 0.2 V.

Fig. 2 the typical electron depletion region (red solid line) under the gate at positive bias while at negative bias the depletion region (green broken line) is absent because the gate is too close to n^+ drain. This depletion region gives the lower current values as compared with drain current at negative bias (see Fig. 1). Figures from 3 to 6 demonstrate various physical quantity profiles in GaN MOSFET

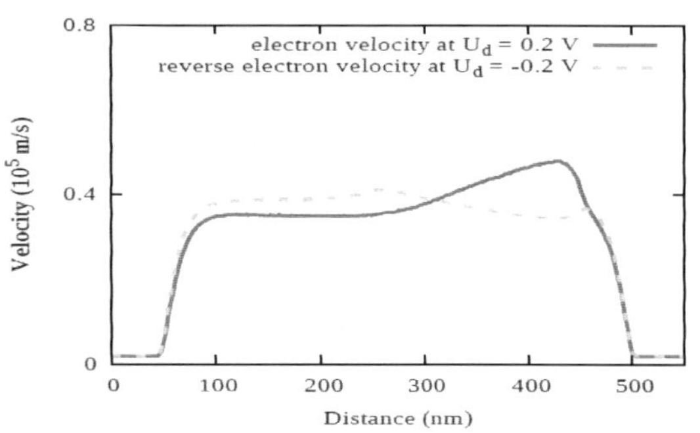

Figure 3. Electron velocity profiles in channel at different drain biases: - 0.2 and 0.2 V.

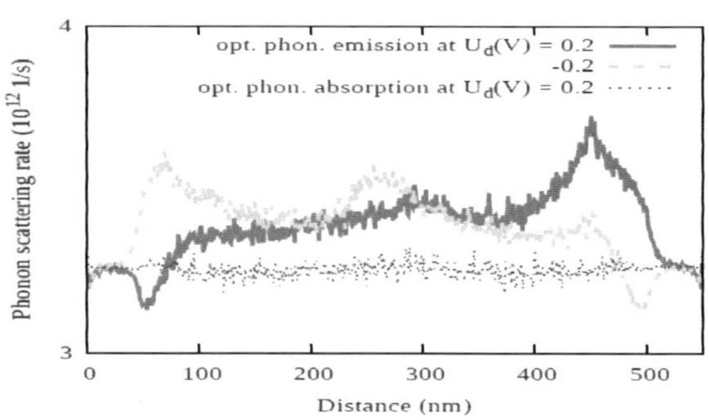

Figure 4. Optical phonon local scattering profiles in channel at different drain biases: -0.2 and 0.2 V.

channel. In Fig. 3 the electron velocity at negative drain bias (green broken line) is shown with negative sign for better comparison with the one at positive drain bias. The increased velocity over the depletion region at positive bias (red solid line) is clearly seen . The optical phonon local scattering profiles are demonstrated in Fig. 4. The optical phonon local absorption profile is flat (Fig. 4, blue dotted line) because the absorption rate do not depend on electron energy. The optical phonon local emission rate profile have two peaks (see Fig. 4). At positive drain bias there is the peak in the channel centre and another one at the drain (Fig. 4, red solid line). At negative drain bias there is the peak in the channel centre and another one at the source (Fig. 4, green broken line). The number of optical phonon local emission peaks depends on potential difference which is flight by electron.

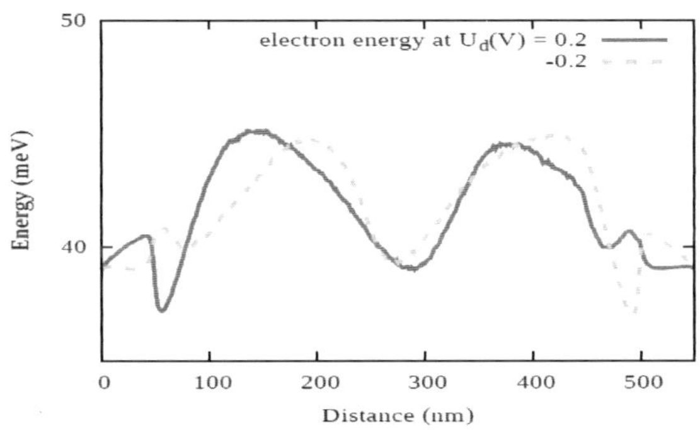

Figure 5. Electron energy profiles in channel at different drain biases: - 0.2 and 0.2 V.

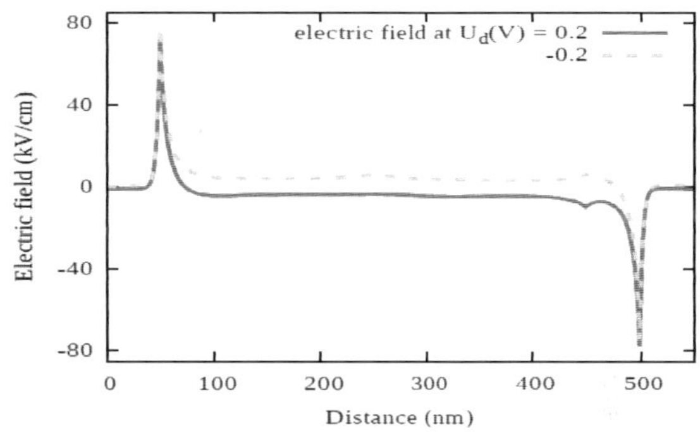

Figure 6. Electric field profiles in channel at different drain biases: - 0.2 and 0.2 V.

Keeping in mind that optical phonon energy in wurtzite GaN is of 91.2 meV the two optical phonon emission local peaks are possible during 0.2 V flight (see Fig.4). The electron energy profiles are demonstrated in Fig. 5. The electron energy minima at the centre of the channel are caused by optical phonon emission (see Fig. 4). Fig. 6. shows the electric field profiles in wurtzite GaN MOSFET channel. The high electric field spikes are formed at n^+n junctions of source and drain contacts.

4. THz radiation detection

The wurtzite GaN MOSFET detector drain current response to the radiation is simulated applying to the structure an alternating bias $U(t) = U_1\cos(2\pi ft)$, where U_1 is voltage induced along the channel by the microwave radiation and f is the radiation frequency. The long current response trajectories are simulated. The drain current value is obtained by averaging in time the stationary part of trajectory.

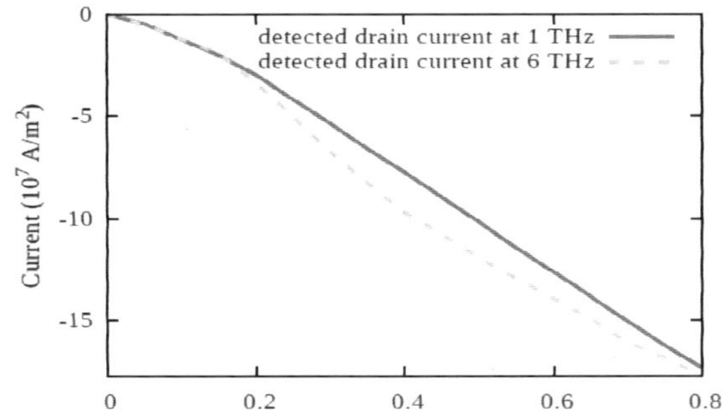

Figure 7. Drain current dependence on alternating drain bias amplitude at frequencies 1 and 6 THz.

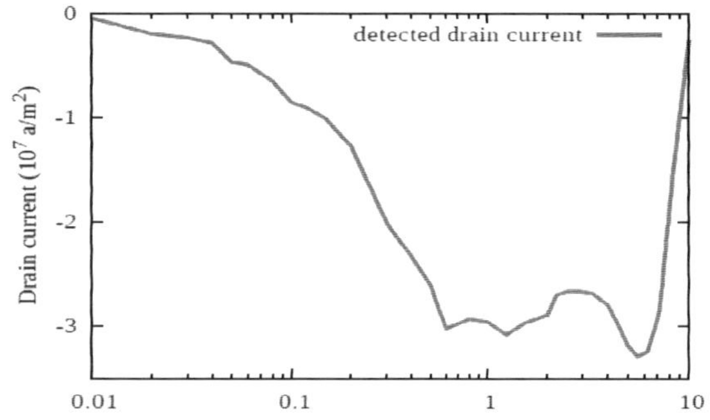

Figure 8. Drain current spectrum at alternating drain bias $U_1 = 0.2$ V.

The drain current dependencies on alternating drain bias amplitude U_1 at frequencies f = 1 and 6 THz and zero drain and gate biases are demonstrated in Fig. 7. The absolute values of detected drain currents monotonically grow with the alternating drain bias amplitude U_1. The detected drain current spectrum in frequency range from 0.01 to 10 THz at alternating drain bias U1 = 0.2 V is presented in Fig. 8. The absolute value of detected drain current monotonically grows with frequency and approach the maximum in frequency range from 0.5 to 7 THz. To explain the frequency dependence of detected drain current the optical phonon local emission profiles in wurtzite GaN MOSFET channel at alternating drain bias $U_1 = 0.2$ V and frequencies f = 0.1 and 1 THz are simulated (see Fig. 9). The considerable increase of optical phonon local emission rate at 1 THz frequency as compared with rate at 0.1 THz is evident. This optical phonon local emission rate resonant increase with frequency f (see

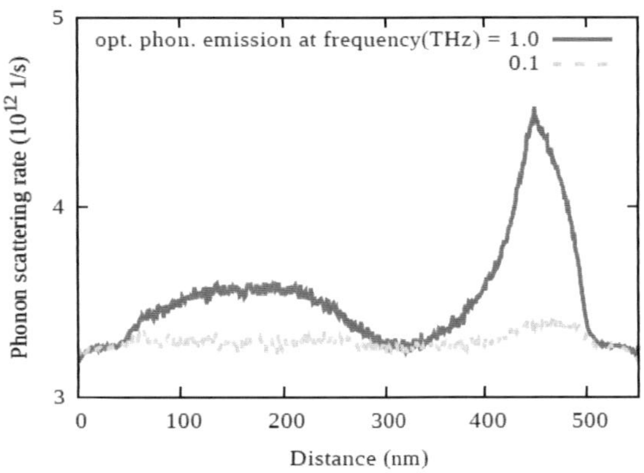

Figure 9. Optical phonon local emission profiles in channel at $U_1 = 0.2$ V, f = 0.1 and 1 THz.

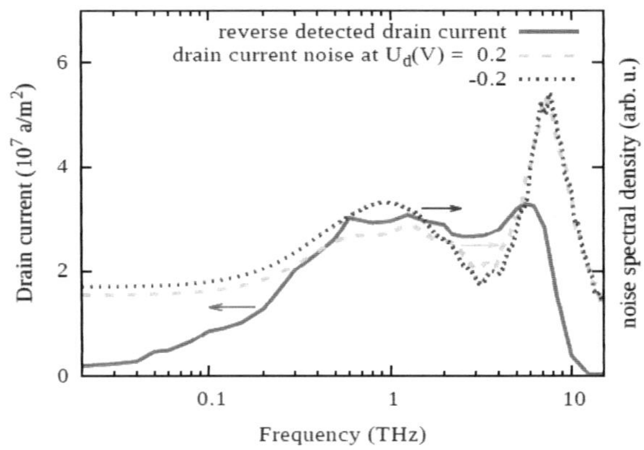

Figure 10. Frequency dependences of detected drain current and drain current noise at $U_d = 0.2$ and -0.2 V.

Fig. 9) gives the increase of detected drain current absolute values (Fig. 8). Also the drain current noise spectral densities are calculated at constant drain biases Ud = 0.2 and -0.2 V (Fig. 10, green broken and blue dotted lines) . The drain current noise spectral densities at negative and positive drain biases have two maxima each. First maximum is around 1 THz and is associated with optical phonon emission induced plasma instability and the second one around 7 THz is the result of plasma oscillations at n^+n junctions of drain and source contacts. The comparison of reverse detected drain current (absolute values of Fig. 8) with spectral noise densities in Fig. 10 clearly show that enhanced detection current in frequency range of 0.5 – 7 THz are coused by optical phonon emission processes. At frequencies over 7 THz the absolute value of detected drain current rapidly decreases due to the inertia of electron motion.

Acknowledgment
This work is supported in part by the Research Council of Lithuania (grant No. LAT-03/2016) in the frame of National Science Program "Towards Future Technologies".

References
[1] A. Sužiedelis, J. Gradauskas,S. Ašmontas, G. Valušis and H. G. Roskos, "Giga- and terahertz frequency band detector based on an asymmetrically necked n – n+-GaAs planar structure", *J. Appl. Phys.*, v. **93**, p.p. 3034-3038, (2003).
[2] V. Gružinskis, E. Starikov, P. Shiktorov and S. Ašmontas, "Monte Carlo simulation of THz radiation detection in GaAs n^+nn^+ diodes with inhomogeneously doped n region", *J. Phys.: Conf. Ser.*, v. **193**, p. 012069, (2009).
[3] G H. Marinchio, C. Palermo, G. Sabatini, L. Varani, P. Shiktorov, E. Starikov, V. Gružinskis, "Pseudo-two-dimensional Poisson equation for the modelling of field-effect transistors". *J. Comput. Electron.*, v.. **9**, p.p. 141-145, (2010).
[4] B. Foutz, S. O'Leary, M. Shur, L. Eastman, "Transient electron transport in wurtzite GaN, InN, and AlN". *J. Appl. Phys.*, v. **85**, p. 7727, (1999).

IOP Conf. Series: Journal of Physics: Conf. Series **906** (2017) 012014 doi:10.1088/1742-6596/906/1/012014

Magnetic-optical transitions induced by twisted light in quantum dots

G. F. Quinteiro[1], D. E. Reiter[2], T. Kuhn[2]

[1]Departamento de Física and IFIBA, FCEN, Universidad de Buenos Aires, Buenos Aires, Argentina

[2]Institut für Festkörpertheorie, Universität Münster, 48149 Münster, Germany

E-mail: gquinteiro@df.uba.ar

Abstract. It has been theoretically predicted that light carrying orbital angular momentum, or twisted light, can be tuned to have a strong magnetic-field component at optical frequencies. We here consider the interaction of these peculiar fields with a semiconductor quantum dot and show that the magnetic interaction results in new types of optical transitions. In particular, a single pulse of such twisted light can drive light-hole-to-conduction band transitions that are cumbersome to produce using conventional Gaussian beams or even twisted light with dominant electric fields.

Introduction

Twisted light (TL) is light having a helical wave front. Such light exhibits several interesting features: Due to the azimuthal phase dependence a phase singularity occurs at the beam axis, leading to the name optical vortex. Furthermore, in addition to spin angular momentum (SAM) associated with the handedness of circular polarization, such light fields carry orbital angular momentum (OAM). Depending on the combination of SAM and OAM, TL beams may have strong field components along the propagation direction or strong magnetic field close to the beam axis. The research in TL spans nowadays several areas of fundamental and applied physics [1] including its potential for quantum communication [2].

Of particular interest is the interaction of a TL beam with matter, which opens up the possibility to address unusual transitions [3, 4] and to excite typically dark modes in plasmonics [5]. In this paper we focus on the interaction of TL with a semiconductor quantum dot (QD). QDs are discussed for many applications in optoelectronics and spintronics [6] which, however, requires a precise control of the optically excited states. Here we will show that the interaction of a QD with TL having a strong magnetic component may be used for the excitation of specific light-hole (LH) excitons, which cannot easily be addressed by conventional Gaussian beams or even twisted light with dominant electric fields.

Twisted light with a strong magnetic component

TL can be characterized by the handedness of its circular polarization σ (SAM) and its OAM quantum number ℓ. We recently identified two distinct classes of TL [3, 4], that we named parallel and antiparallel class reflecting the relative orientation of SAM and OAM. Figure 1 depicts the electric field profiles for a beam with $\ell = 2$ and $\sigma = \mp 1$ showing the pronounced difference in the topology of the field lines.

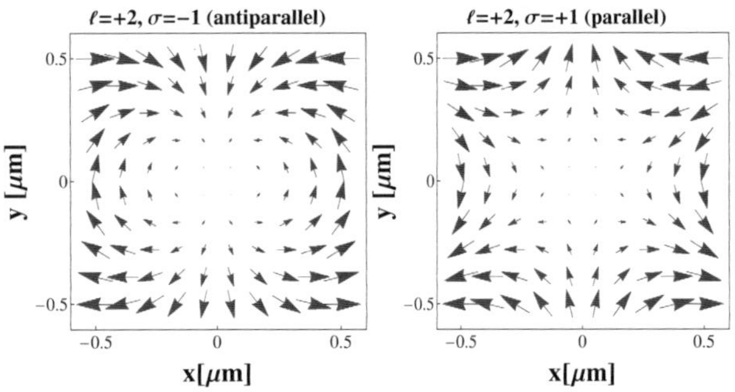

Figure 1. Electric field patterns for the antiparallel (left) and parallel (right) class for an orbital angular momentum of $\ell = 2$.

Starting from the vector potential for a Bessel beam with given values of ℓ and σ propagating in z-direction, which is an exact solution of the vectorial Helmholtz equation, the behavior of the electric and magnetic field in a region close to the beam axis has been derived in [3]. Remarkably, it has been found that especially the antiparallel class is characterized by unconventional features which become particularly strong in the case of tightly focused beams: (i) The electric field close to the beam axis is dominated by its longitudinal component; (ii) for $|\ell| \geq 2$ the beam in that region is dominated by the magnetic field. The virtue of employing the longitudinal electric field component of antiparallel beams with $|\ell| = 1$ for the excitation of LH excitons has already been discussed in [7]. Here we complement this work by showing that the use of the dominant magnetic field present in an antiparallel beam with $|\ell| = 2$ again opens up new excitation pathways which are usually unaccessible.

Close to the singularity the field amplitudes read (see Table 1 in [3])

$$\tilde{E}_x(\mathbf{r}_\perp) = i\frac{\tilde{E}_0}{8}(q_r r)^2 e^{i2\varphi}\,, \qquad \tilde{B}_x(\mathbf{r}_\perp) = -\frac{\tilde{B}_0}{2}\left(\frac{q_r}{q_z}\right)^2,$$

$$\tilde{E}_y(\mathbf{r}_\perp) = \frac{\tilde{E}_0}{8}(q_r r)^2 e^{i2\varphi}\,, \qquad \tilde{B}_y(\mathbf{r}_\perp) = -i\frac{\tilde{B}_0}{2}\left(\frac{q_r}{q_z}\right)^2,$$

$$\tilde{E}_z(\mathbf{r}_\perp) = -\frac{\tilde{E}_0}{2}\frac{q_r}{q_z}(q_r r)\, e^{i\varphi}\,, \qquad \tilde{B}_z(\mathbf{r}_\perp) = -i\frac{\tilde{B}_0}{2}\frac{q_r}{q_z}(q_r r)e^{i\varphi}\,.$$

Here, q_r and q_z are the transverse and longitudinal wave vector, where q_r is inversely proportional to the beam waist, $\mathbf{r}_\perp$ is the in-plane coordinate with $|\mathbf{r}_\perp| = r$ and azimuthal angle φ. We will consider the excitation by tightly focused TL, i.e., beams with the paraxial parameter $q_r/q_z \approx 1$. $\tilde{E}_0$ and $\tilde{B}_0 = \tilde{E}_0/c$ are the respective amplitudes.

When approaching the beam center, i.e. for $r \to 0$, we find that the electric field $\mathbf{E}$ and the longitudinal magnetic field B_z vanish while the transverse magnetic field $\mathbf{B}_\perp$ remains finite. Therefore, the transverse magnetic field dominates for interaction with particles or nanostructures, which are much smaller than the beam waist and are placed at or close to $r = 0$, which is in fact a surprising effect at optical frequencies.

Recently, we have developed the theory of light-nanostructure interaction for the electric and magnetic components of general antiparallel TL beams [3, 4]. It turned out that the interaction can be written in a form similar to the well-known dipole approximation for electric and magnetic fields [4]. In particular, for TL with $\ell = 2$ and $\sigma = -1$, in which the magnetic field dominates, the Hamiltonian can be written as [4]

$$H_I = -\frac{q}{2m}\mathbf{B}_\perp(t)\cdot(\mathbf{r}\times\mathbf{p}) = -i\frac{q\tilde{B}_0}{m}\,(r_+\,p_z - z\,p_+)\,e^{-i\omega t} + \text{c.c.} \tag{1}$$

with q and m being the electron charge and mass, respectively, ω is the frequency of the light, $r_\pm = x \pm iy$, $p_\pm = p_x \pm ip_y$, and c.c. denoting the complex conjugate.

Quantum dot light-hole states

We consider a semiconductor QD, where the conduction band states are s-type ($|s\rangle$) and can be classified by the electron spin given by the quantum number $s_z = \pm 1/2$. For the valence band states ($|p_x\rangle, |p_y\rangle, |p_z\rangle$), due to their p-type character, one finds states with total orbital angular momenta $j = 3/2$ and $j = 1/2$, where the $j = 3/2$ states are split into heavy holes with $j_z = \pm 3/2$ and LHs with $j_z = \pm 1/2$. Here, we focus in the LH states, such that the Bloch function for electrons and holes read:

$$| s_z = +1/2\rangle = |s\rangle\,|\uparrow\rangle \qquad | j_z = +1/2\rangle = \frac{1}{\sqrt{6}}[(| p_x\rangle - i\,| p_y\rangle)\,|\uparrow\rangle + 2\,| p_z\rangle\,|\downarrow\rangle]$$

$$| s_z = -1/2\rangle = | s\rangle\,|\downarrow\rangle \qquad | j_z = -1/2\rangle = -\frac{1}{\sqrt{6}}[(| p_x\rangle + i\,| p_y\rangle)\,|\downarrow\rangle - 2\,| p_z\rangle\,|\uparrow\rangle]$$

where the arrow indicates the electron spin. Note that the sign of j_z refers to the angular momentum of the holes which is opposite to that of valence band electrons.

To describe the full wave function of electrons and holes in the QD, we further apply the envelope function approximation. Within this scheme the Bloch functions are multiplied by the respective envelope functions for electrons and holes,

$$\mathcal{F}^{e/h}_{n,m}(\mathbf{r}) = \mathcal{R}^{e/h}_{n,m}(r)\, e^{im\varphi}\, \mathcal{Z}^{e/h}(z)\ , \tag{2}$$

which have been further separated into the radial part $\mathcal{R}^{e/h}_{n,m}(r)$, an envelope angular momentum part $e^{im\varphi}$, and a part $\mathcal{Z}^{e/h}(z)$ in z-direction. Assuming a flat QD, we restrict ourselves to the ground state in z-direction assuming wave functions $\mathcal{Z}^{e/h}$ with well-defined parity.

The interaction of twisted light with light holes

To calculate the matrix elements between LH valence and conduction band states we follow the standard procedure in the envelope-function formalism: We split the integration into a sum over all lattice vectors $\mathbf{R}$ and an integral over the unit cell indicated by the coordinate $\mathbf{r}'$ [8]. Finally, we replace the sum over lattice vectors by an integral over the whole system.

Following this procedure, in the Hamiltonian (1) the products of coordinates and momenta separate into terms such as $(R_+ + r'_+)(P_z + p'_z)$. Keeping in mind the orthogonality and parity of the Bloch functions under study, one finds that all matrix elements involving only intracell coordinates $\mathbf{r}'$ or only envelope coordinates $\mathbf{R}$ vanish. Furthermore, the integral over the envelope functions including either Z or P_z vanish, i.e., $\int dZ\, \mathcal{Z}^{e*}(Z)\, Z\, \mathcal{Z}^h(Z) = \int dZ\, \mathcal{Z}^{e*}(Z)\, P_z\, \mathcal{Z}^h(Z) = 0$, due to the same parity of the ground state functions $\mathcal{Z}^{e/h}(z)$. The only remaining matrix element then reads

$$\langle n', m';\, s_z \mid H_I \mid n, m;\, j_z\rangle = -i\frac{q\tilde{B}_0}{m}\int d^3R\, \mathcal{F}^{e*}_{n',m'}(\mathbf{R})\, [R_+ M_{p_z} - P_+ M_z]\, \mathcal{F}^h_{n,m}(\mathbf{R})\delta_{s_z,-j_z}\ , \tag{3}$$

where we have introduced the microscopic matrix elements

$$M_z = \langle s_z = \pm 1/2|\, z'\, |j_z = \mp 1/2\rangle \qquad M_{p_z} = \langle s_z = \pm 1/2|\, p'_z\, |j_z = \mp 1/2\rangle\ .$$

In Eq. (3), we find that due to the optical selection rules only electron-hole pairs having opposite angular momenta of the Bloch state are excited, because only the matrix element M_z and M_{p_z} survive. In other words, only excitons with a total angular momentum of zero (i.e. $J_z = s_z + j_z = 0$) are excited. This is very different compared to the excitation with plane waves. Plane waves have an angular momentum of ± 1 and thereby excite excitons with a total angular momentum $J_z = \pm 1$. To excite LH excitons in a QD with $J_z = 0$, one would have to apply an excitation from the side of the QD, which typically requires cleaving the sample.

We can now evaluate the macroscopic integrals over R for the two different parts. For the first term, we rewrite $R_+ = Re^{i\Phi}$. For the second term we write the momentum operator in cylindrical coordinates $P_+ = -i\hbar e^{i\Phi}\left[\partial_R + (i/R)\partial_\Phi\right]$. Thus, both terms contribute a term $\sim e^{i\Phi}$ to the angle integral. This shows, that by exciting the QD with a magnetic TL beam having $\ell = 2$ and $\sigma = -1$, the envelope state changes by $\Delta m = 1$ while the total band+spin angular momentum of the exciton is zero. Such a transition is forbidden for excitation with a plane wave due to dipole selection rules, and it is also not present in the case of excitation by a TL beam with dominant electric field in z-direction (i.e., an antiparallel TL beam with $|\ell| = 1$) [7].

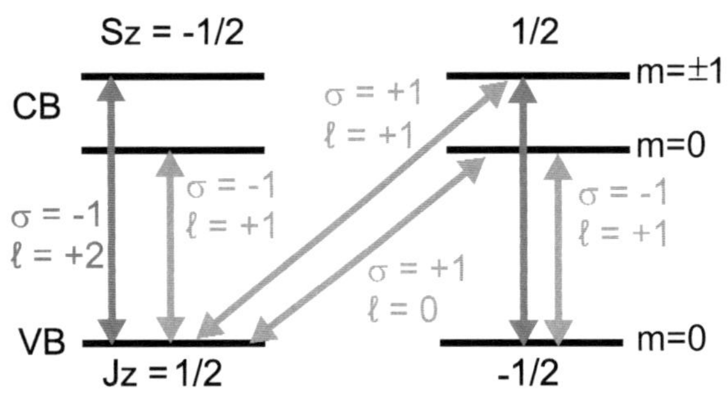

Figure 2. Optical transition between light-hole and conduction band state in a QD induced by single pulses of twisted light having $\ell = 2$ and $\sigma = -1$. For comparison also transition due to beams with $\ell = 0$ and $\ell = 1$ are included [7]. Note that for clarity reasons holes states with $m = \pm 1$ have been removed.

Figure 2 combines our new results for a TL beam with $\ell = 2$ and $\sigma = -1$ with previous results reported in Ref. [7] for beams with $\ell = 0$ and $\ell = 1$. Note that for reasons of clarity the Figure shows only the lowest hole shell with $m = 0$. The diagonal transitions with $\Delta m = 0$ and $\Delta m = -1$ induced correspondingly by TL beams with $(\sigma = -1, \ell = 0)$ and $(\sigma = -1, \ell = -1)$ are not depicted. We thus find, that by using TL beams with $|\ell| \leq 2$ all possible exciton states in the lowest two shells with $m = 0$ and $m = \pm 1$ can be excited.

Conclusions

We have demonstrated that the optical-magnetic interaction in case of excitation with a TL beam with OAM $\ell = 2$ and SAM $\sigma = -1$ generates LH exciton states with no band+spin angular momentum ($J_z = 0$) but with envelope angular momentum $m = 1$, and thus produces states that are not accessible with plane waves or even TL with dominant electric interaction. Combining these findings with previous results for beams with other values of OAM and SAM, we find that TL beams are beneficial to obtain a full control of exciton states in a QD.

References

[1] Andrews D L 2008 *Structured light and its applications: An introduction to phase-structured beams and nanoscale optical forces* (Academic Press)

[2] Mair A, Vaziri A, Weihs G and Zeilinger A 2001 *Nature* **412** 313

[3] Quinteiro G F, Reiter D E and Kuhn T 2015 *Phys. Rev.* A **91** 033808

[4] Quinteiro G F, Reiter D E and Kuhn T 2017 *Phys. Rev.* A **95** 012106

[5] Kerber R M, Fitzgerald J M, Reiter D E, Oh S S and Hess O 2017 *ACS Photon.* **4** 891

[6] Chen P, Piermarocchi C, Sham L J, Gammon D and Steel D G 2004 *Phys. Rev.* B **69** 075320

[7] Quinteiro G F and Kuhn T 2014 *Phys. Rev.* B **90** 115401

[8] Haug H and Koch S W 2004 *Quantum Theory of the Optical and Electronic Properties of Semiconductors* 4th ed (Singapore: World Scientific)

Study for Enhancement of Terahertz Radiation Using GaSb/InAs Heterostructures

S Sasa[1], Y Kinoshita[1], M Tatsumi[1], M Koyama[1], T Maemoto[1], S Hamauchi[2], I Kawayama[2] and M Tonouchi[2]

[1] Nanomaterials and Microdevices Research Center, Osaka Institute of Technology, 5-16-1 Ohmiya, Asahi-ku, Osaka 535-8585, Japan
[2] Institute of Laser Energy, Osaka University, 2-6 Yamadaoka, Suita 565-0871, Japan

E-mail: shigehiko.sasa@oit.ac.jp

Abstract. GaSb/InAs heterostructures were investigated for the enhancement of terahertz (THz) radiation from an InAs thin film excited by a femtosecond laser. Taking advantage of the large conduction band discontinuity between GaSb and InAs, electrons excited in the GaSb layer gain a large excess energy when injected into the InAs radiation layer. Enhancement of the THz radiation by a factor of approximately two was observed when the surface pinning position was controlled to reduce the surface electric field by the introduction of a thin InAs cap layer.

1. Introduction

Methods for generating terahertz (THz) radiation have been extensively studied because of growing interest in applications of this type of radiation [1]. THz radiation emitted from semiconductor surfaces excited by femtosecond laser pulses is of great practical importance. InAs is one of the best studied III-V compound semiconductors because it emits intense THz radiation [2-8]. We have previously reported [9] that InAs thin films with thicknesses of 1 μm or greater produce THz radiation that is more intense than that from a p-type InAs substrate. The use of InAs thin films grown on low-cost substrates such as GaAs is advantageous for the development of THz emitting devices. Furthermore, we have reported that a heavily-doped InAs layer under the InAs radiation layer acts as a plasma reflection layer that leads to a further enhancement in THz emission [10]. In the present study, we investigated the enhancement of terahertz radiation from an InAs thin film by the introduction of a GaSb hot-electron injector to strengthen the surge current. The large conduction band discontinuity between GaSb and InAs means that we can expect the electrons injected from the GaSb injector into the InAs radiation layer to gain excess energy that enhances the surge current, and thus the THz radiation intensity. We also demonstrate that control of the surface pinning effect is critical for enhanced THz radiation.

2. Experimental

Epitaxial GaSb/InAs heterostructures were grown on a semi-insulating GaAs(100) substrate by molecular beam epitaxy. Growth was initiated with a 100 nm thick InAs low-temperature buffer layer grown at 300°C. This buffer layer was used to replace the thicker Sb-based buffer layer employed in previous experiments [1] while maintaining a high electron mobility of about 7000

IOP Conf. Series: Journal of Physics: Conf. Series **906** (2017) 012015 doi:10.1088/1742-6596/906/1/012015

cm^2/Vs at room temperature. A 1 μm thick InAs radiation layer and a GaSb injection layer with thicknesses from 0 to 300 nm were then successively grown at 450°C. The GaSb thickness range was chosen to cover the absorption length of 190 nm for the wavelength of the laser pump pulses [11]. The growth rate for both InAs and GaSb was 0.6 μm/h. Figure 1 shows a cross-sectional schematic of the GaSb/InAs heterostructure.

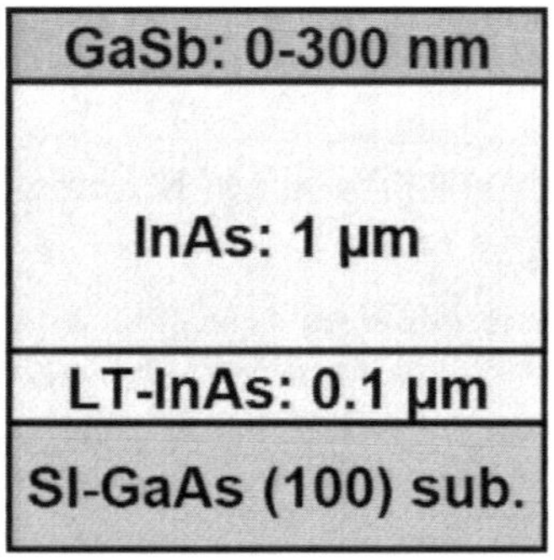

Fig. 1. Cross-sectional illustration of GaSb/InAs heterostructure. The thickness of the GaSb layer was varied between 0 and 300 nm.

Femtosecond laser pulses produced by a mode-locked Ti:sapphire laser were used as the optical excitation. The excitation wavelength was 800 nm, the pulse width was 100 fs, and the pulse repetition rate was 82 MHz. The laser power was 80 mW and the laser pulses were focused to a spot diameter of about 5 mm on the sample surface with an incidence angle of 45° from the surface normal. Time-domain THz spectroscopy was used to observe the waveforms from the sample using a reflection geometry. Since a relatively large spot size of 5 mm was used for the excitation, the emission spectrum showed a peak at around 0.5 THz, which is lower than that in previous experiments [1]. The relative emission intensity for an InAs thin film, an n-InAs substrate, and a p-InAs substrate was confirmed to be comparable with that obtained previously, despite the fact that the excitation conditions were different. Figure 2 shows the dependence of the emission intensity for

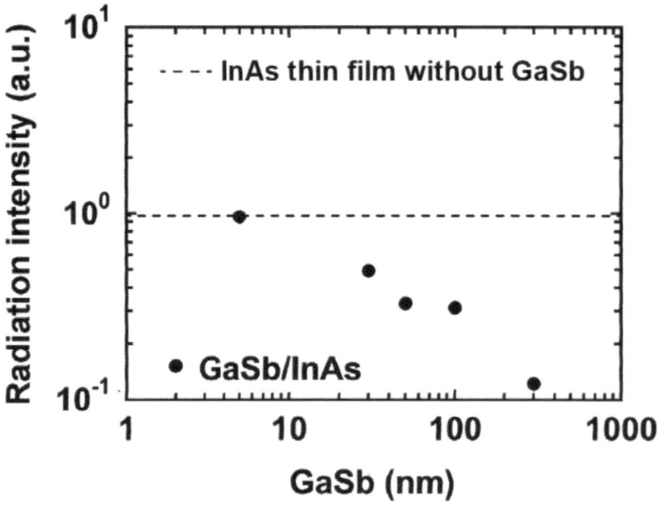

Fig. 2. Terahertz emission intensity for GaSb/InAs heterostructure as function of GaSb thickness. The intensity is normalized with respect to that for a 1 μm thick InAs thin film.

IOP Conf. Series: Journal of Physics: Conf. Series **906** (2017) 012015 doi:10.1088/1742-6596/906/1/012015

the GaSb/InAs heterostructure on the GaSb injection layer thickness. The intensity is normalized with respect to that for a 1 μm thick InAs thin film without GaSb, indicated by the dashed line. The radiation intensity increases as the thickness of the GaSb injection layer decreases from 300 nm to 5 nm. However, the intensity is weaker than that for the InAs thin film. This may be due to Fermi-level pinning at the GaSb surface or inter-valley scattering in the InAs emission layer. The surface pinning level for GaSb is reported to be 0.45 eV below the bottom of the conduction band edge [12]. Therefore, a strong electric field is produced along the growth axis and suppresses electron diffusion toward the substrate. The surface pinning effect was also confirmed by observation of Shubnikov-de Haas oscillations, which indicate the formation of a two-dimensional electron gas at the GaSb/InAs interface as a result of the strong surface electric field. In GaSb/InAs heterostructures, excited electrons are expected to be injected into higher energy levels than in the case of InAs thin films, and such electrons may undergo frequent inter-valley scattering. Inter-valley scattering of electrons is detrimental, because this greatly reduces their mobility, leading to a reduction in the diffusion current and thus the radiation intensity.

A GaSb/InAs heterostructure with a 1 nm thick InAs cap layer was then grown to examine the surface pinning effect. The GaSb injection layer in this structure has a thickness of 50 nm and is sandwiched by n-type InAs layers, so that the conduction band of the GaSb levels off to reduce the electric field. Figure 3 shows the time variation of the radiation intensity for GaSb/InAs heterostructures with and without an InAs cap layer. The intensity for the structure with the InAs cap layer is approximately two times higher than that for the structure without the InAs cap layer. This strongly suggests that the surface pinning effect for the GaSb/InAs heterostructure hinders the enhancement of THz emission, whereas the introduction of the thin InAs cap layer recovers the emission intensity.

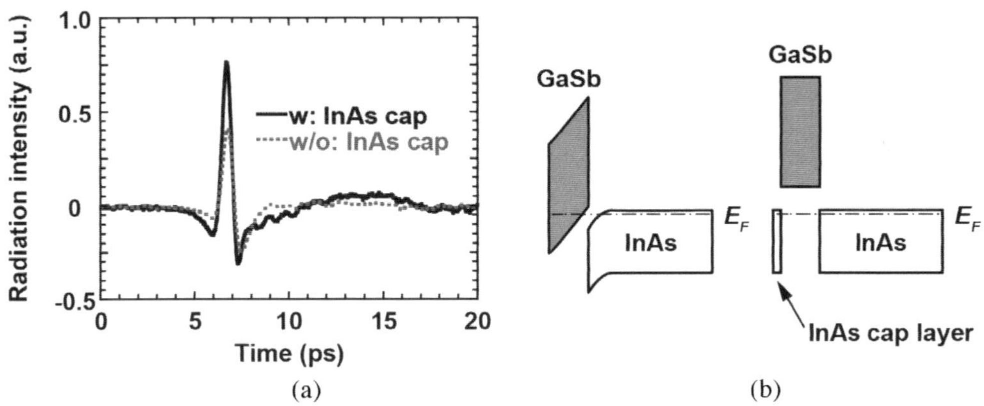

Fig. 3. (a) Time dependence of terahertz emission intensity for GaSb/InAs and InAs/GaSb/InAs heterostructures. Enhanced radiation is observed for the InAs/GaSb/InAs heterostructure. (b) Band diagrams for GaSb/InAs heterostructures without an InAs cap layer (left) and with an InAs cap layer (right).

Conclusion

THz radiation from an InAs thin film was investigated using a GaSb hot-electron injector to enhance electron diffusion, and thus the THz radiation intensity. The intensity for a GaSb/InAs heterostructure was lower than that for an InAs thin-film structure due to strong surface band bending in the GaSb/InAs heterostructure. The introduction of a thin InAs cap layer increased the radiation intensity by a factor of two. This constitutes evidence that control of surface pinning is critical for achieving enhanced THz radiation from GaSb/InAs heterostructures.

IOP Conf. Series: Journal of Physics: Conf. Series **906** (2017) 012015 doi:10.1088/1742-6596/906/1/012015

Acknowledgments
This work was supported in part by a Kakenhi Grant-in-Aid (No. 16K06326) from the Japan Society for the Promotion of Science (JSPS).

References
[1] D. H. Auston, K. P. Cheung, and P. R. Smith, Appl. Phys. Lett. 45, 284 (1984).
[2] M. Suzuki, M. Tonouchi, K. Fujii, H. Ohtake, and T. Hirosumi, Appl. Phys. Lett. 89, 091111 (2006).
[3] P. Gu, M. Tani, S. Kono, K. Sakai, and X.-C. Zhang, J. Appl. Phys. 91, 5533 (2002).
[4] R. Adomavičius, G. Molis, A. Krotkus, and V. Sirutkaitis, Appl. Phys. Lett. 87, 261101 (2005).
[5] H. Takahashi, A. Quema, M. Goto, S. Ono, and N. Sarukura, Jpn. J. Appl. Phys. 42, 1259 (2003).
[6] R. Adomavičius, A. Urbanowicz, G. Molis, A. Krotkus, and E. Šatkovskis, Appl. Phys. Lett. 85, 2463 (2004).
[7] K. Liu, J. Xu, T. Yuan, and X.-C. Zhang, Phys. Rev. B 73, 155330 (2006).
[8] R. Inoue, K. Takayama, and M. Tonouchi, J. Opt. Soc. Am. B 26, A14 (2009).
[9] S. Umino, Y. Ishibashi, T. Maemoto, S. Sasa, and M. Inoue, K. Takeya, and M. Tonouchi, Extended Abstracts of 2nd Int. Workshop on Terahertz Technology (TeraTech '09), 205 (2009).
[10] Michał Kozub. Kazuichi Nishisaka, Toshihiko Maemoto, Shigehiko Sasa, Kazuhisa Takayama Masayoshi Tonouchi, J Infrared Milli Terahz Waves, 36, 423 (2015).
[11] D. E. Aspnes and A. A. Studna, Phys. Rev. B 27, 985 (1983).
[12] C. Nguyen, B. Brar, and H. Kroemer, J. Vac. Sci. Technol. B 11, 1706 (1993).

IOP Conf. Series: Journal of Physics: Conf. Series **906** (2017) 012016 doi:10.1088/1742-6596/906/1/012016

Characterization of (Cd,Mn)Te and (Cd,Mg)Te single crystals in the THz frequency range using integrated photoconductive and electro-optic effects

J Serafini,[1] S B Trivedi,[2] D Kochanowska,[3] M Witkowska-Baran,[3] A Mycielski,[3] M Guziewicz,[4] R Kruszka,[4] W Słysz,[4] and Roman Sobolewski[1]

[1]University of Rochester, Rochester, NY 14627-0231, USA
[2]Brimrose Technology Corporation, Sparks, MD 21152, USA
[3]Institute of Physics, Polish Academy of Sciences, Warszawa, PL-02668 Poland
[4]Institute of Electron Technology, Warszawa, PL-02668 Poland

E-mail: roman.sobolewski@rochester.edu

Abstract. We present THz frequency range characterization of highly resistive (Cd,Mg)Te and (Cd,Mn)Te single crystals, using an "experiment-on-chip" configuration. We have demonstrated that both of these single crystals exhibit simultaneously strong photoconductive (PC) and electro-optic (EO) effects by performing measurements on a given single platelet with a deposited Au coplanar transmission line. We optically generated a subpicosecond electrical transient by focusing an ultraviolet 100-fs-wide pump pulse between the electrodes of a dc-biased coplanar line (PC effect) and, subsequently, time resolved it with a subpicosecond resolution along the transmission line using an internal EO effect by passing infrared, 100-fs-wide probe pulses through the crystal between the coplanar strips. Transients sampled at different distances from the generation site allowed us to calculate the complex propagation factor $\gamma(f)$ of our transmission lines and the corresponding THz bandwidth attenuation and phase velocity. The latter parameters enabled us to reconstruct the original ~600-fs-in-duration, PC-generated transient by "back-propagating" a signal to the excitation point. Furthermore, we have also determined the THz-bandwidth EO coefficients of our (Cd,Mn)Te and (Cd,Mg)Te crystals to be 6 pm/V and 1.2 pm/V, respectively.

1. Introduction

Ternary compounds based on the CdTe crystal structure, such as $Cd_{1-x}Mg_xTe$ [(Cd,Mg)Te, (CMgT)], $Cd_{1-x}Zn_xTe$ [(Cd,Zn)Te, (CZT)], and $Cd_{1-x}Mn_xTe$ [(Cd,Mn)Te, CMnT] have recently been receiving widespread research attention because of their interesting physical properties with broad device applications, ranging from photoconducting and optoelectronic to magneto-optic. The main advantage of ternary compounds over plain CdTe single crystals is that incorporation of the third element leads to an enhancement of their energy bandgap E_G with only minimal deterioration of either crystalline or transport (carrier mobility) parameters. The increase of the value "x" leads to roughly a linear increase of E_G, which optimizes the device's performance. They are also very highly resistive materials, and with additional carrier compensation using, e.g., shallow trap states can reach resistivities above 10^{10} Ω-cm. While CZT has become a leading material for solid-state x- and gamma-ray detectors [1], CMnT single crystals are the most-studied diluted magnetic semiconductors [2]. They possess unique

IOP Conf. Series: Journal of Physics: Conf. Series **906** (2017) 012016 doi:10.1088/1742-6596/906/1/012016

magneto- and electro-optic properties [3,4], and most recently have been successfully implemented in time-resolved x-ray detectors [5]. (Cd,Mg)Te is the most-recent extension of the CdTe-based ternary family [6] and exhibits the widest $E_G(x)$ tuning range, which extends from the near infrared to the visible part of the spectrum.

2. Sample fabrication and experimental setup

Our CMgT and CMnT samples were grown using a vertical Bridgman method, following procedures described in [7] and [8], respectively. After the growth, the crystals were cut into ~50-mm^2-area, 1-mm-thick platelets. CMnT crystals were also annealed in Cd vapour to improve their crystallinity and reduce the number of Cd vacancies, while CMgT crystals were Ge doped. In both cases, their resulting resistivities were at least at the 10^6 Ω-cm level. A 200-nm-thick Au film was sputtered on the top (110) surface of the samples; subsequently, coplanar strip (CPS) transmission lines were fabricated using optical lithography and a lift-off process. Fabricated CPS lines were 100 μm wide with a 25-μm gap, as shown in the inset in figure 1.

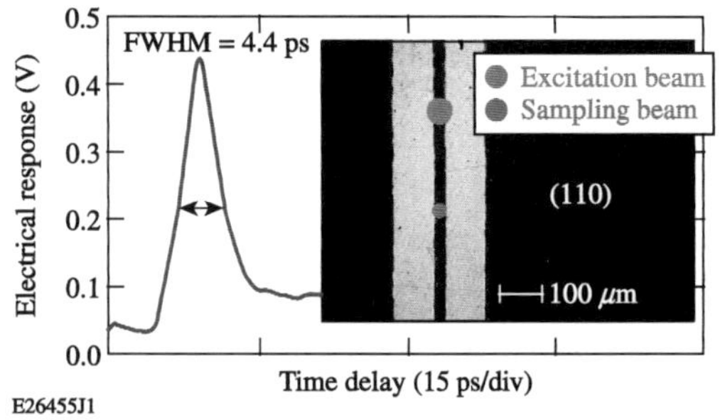

Figure 1: Time-resolved PC response of a CMnT crystal with the inset featuring our "experiment-on-chip" configuration, with excitation and sampling beams, blue and red dots, respectively. The excitation pulse had a pump power of 10 mW with a 415-nm wavelength. A 10-V dc bias was applied to the CPS lines.

Time-resolved measurements of the photoconductive (PC) response using the electro-optic (EO) sampling technique were performed using an "experiment-on-chip" configuration described in detail in [9]. Briefly, we used a mode-locked Ti:sapphire laser emitting a train of ~100-fs pulses at 830 nm at a 76-MHz repetition. The output was split, using a 60/40 beam splitter, into a sampling beam and an excitation beam, the latter of which was frequency doubled using a BaB_2O_4 crystal. The excitation pulses (blue light, 415 nm) were focused to an ~50-μm-diam spot to uniformly illuminate the gap between the dc-biased CPS lines, forming a so-called PC sliding contact [10] (see the inset in figure 1), and triggering surface-photo–generated electrical transients, which were launched into the CPS line. The sampling pulses (infrared light, 830 nm) were focused to a 25-μm-diam spot and passed through the gap between the CPS lines. In the presence of an electric field from a propagating electrical transient, their polarization plane was tilted by δ degrees because of the Pockels (EO) effect, which for the F-43m symmetry group can be expressed as [11]

$$\delta = \frac{2\pi n^3(\lambda) r_{41} V L}{\lambda d}, \tag{1}$$

where $n(\lambda)$ is the refractive index at the wavelength λ of the probe beam, V is the applied voltage, d is the CPS electrode separation distance, r_{41} is the EO coefficient, and L is the interaction length, i.e., penetration of the electric field into the crystal. On the other side of the sample, the probe beam was fed into a polarizing beam splitter and, finally, detected by a pair of balanced photodetectors and a lock-in amplifier at the modulation frequency of the pump pulse train.

The system response was calibrated by measuring the voltage response (amount of birefringence) induced by a continuous-wave sinusoidal signal with a 1.16-V amplitude and a 243-kHz frequency applied between the CPS lines under a standard assumption that the EO effect is not sensitive to the electric-field frequency so long as the energy of the probing photons is below E_G of an EO crystal [10].

3. Experimental results and discussion

We collected a number of CMgT and CMnT photoresponse transients at several distances away from the excitation point, ranging from 100 to 300 μm. Figure 1 features a transient recorded 150 μm from the excitation point for CMnT. The signal exhibits a 3.5-ps rise time (based on the 10% to 90% amplitude criterion), while its trailing edge represents the photocarrier relaxation dynamics and can be very well fitted by a single exponential function with a decay time of 3.7 ps. A slightly wider transient was recorded for the CMgT crystal (not shown).

To determine the dielectric properties of both of our crystals at THz frequencies and the amount of distortion introduced by our CPS line, we performed a standard frequency-domain analysis of our propagated transients [9] and computed frequency-dependent complex propagation factor $\gamma(f)$ for our transmission lines by dividing the frequency spectra of the waveforms measured at points 150 μm and 250 μm from the excitation spot. Figure 2a presents spectral dependencies of the attenuation constant $\alpha(f)$ [real part of $\gamma(f)$] and the transient phase velocity $v(f)$ for CMnT (solid lines) and CMgT (dashed lines), respectively. We note that in both cases $\alpha(f)$ has a rather low value (~4 mm^{-1} and ~6 mm^{-1}, respectively) and remains constant for up to ~0.6 THz for CMnT and ~0.8 THz for CMgT. At highest frequencies, $\alpha(f)$ increases very rapidly because of the expected strong increase in the radiative loss of the CPS line [12]. The phase velocity remains essentially flat with an average $v(f)$ value close to 1.3×10^8 m/s, so we can conclude that our CPS lines can be regarded as the quasi-TEM (transverse electromagnetic) type.

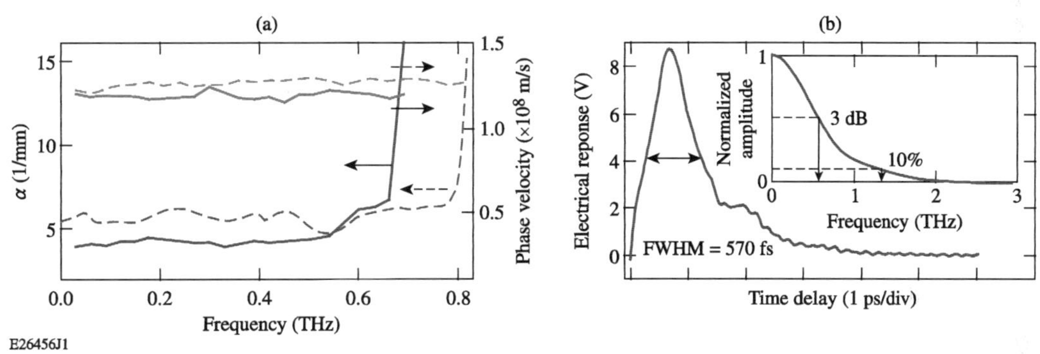

Figure 2: (a) Spectral dependencies of the attenuation constant [$\alpha(f)$] and the transient phase velocity for CMnT (solid lines) and CMgT (dashed lines). (b) The intrinsic PC transient of a CMnT crystal acquired by back-propagating the transient sampled at 150 μm from the photoswitch. The inset shows the power spectrum of the intrinsic pulse.

Knowing $\gamma(f)$, we numerically back-propagated one of the experimental transients toward its zero-distance plane (excitation spot) on the CPS line. The calculated intrinsic transient at the photodetector point for CMnT is shown in figure 2b. We see that it has a peak amplitude of 870 mV, a full width at half maximum (FWHM) of 570 fs, and a fall time of ~700 fs. Note that our intrinsic photoresponse signal has a 3-dB bandwidth up to 560 GHz and a 10% bandwidth above 1.4 THz (figure 2b inset). The signal's subpicosecond fall time indicates that our CMnT crystal must have a large number of shallow traps, and that trapping is the main early-stage carrier relaxation mechanism. Not shown are similar results for the CMgT crystal: an intrinsic pulse had an amplitude of 750 mV, a FWHM of 620 fs, and 3-dB and 10% bandwidths ~ 480 GHz and above 1.2 THz, respectively.

In our experiment-on-chip configuration, the crystals themselves act as EO intensity modulators and measure the change in probe intensity that, subsequently, is recorded by the photodetector and read out by the lock-in as the voltage ΔV. Using the differential transfer function of our EO sampling system [9], we can determine the optically induced phase retardance δ and calculate r_{41} from Eq. 1. As a reference point, we used an averaged pump power of 10 mW. Under these conditions, ΔV = 150

μV and 30 μV for CMnT and CMgT, respectively, at a distance of 150 μV from the photoswitch. Ellipsometry measurements were performed and determined that $n(\lambda = 830\text{ nm}) = 3.0$ for both samples. The field interaction length L in Eq. 1 was taken as the inverse of $\alpha(f)$ for moderate frequencies (see figure 2a). Using the above values, as well as the peak amplitudes of our electrical transients (figure 1) and the CPS dimensions, we calculated r_{41} to be ~6 and ~1.2 pm/V for CMnT and CMgT, respectively, well within the range of EO coefficients reported in the literature [14].

4. Conclusions

We have demonstrated simultaneous subpicosecond PC and EO effects, measured in the same CMgT and CMnT single-crystal samples with a separation distance between PC excitation and EO sampling as short as 150 μm. By sampling the PC transients at different points along the transmission line, we managed to fully characterize propagation of our signals along the CPS line at the sub-THz frequency range, specifically the phase velocity and the attenuation constant. By back-propagating our experimental waveforms, we obtained the intrinsic PC transients, generated using 100-fs-wide, 415-nm-wavelength optical pulses. For CMnT, e.g., such transient exhibited for a 570-fs pulse width with an amplitude of over 700 mV The observed subpicosecond PC response demonstrates that these crystals should function well as ultrafast radiation detectors. Finally, we were able to calculate THz-frequency r_{41} coefficients for both of our CMnT and CMgT crystals.

Acknowledgements

This work was supported in part by the PumpPrimerII Program at the University of Rochester, and by the National Science Centre, Poland, through grant 014/13/B/ST3/04423 and the European Union within the European Regional Development Fund POIG.01.01.02-00-108/09.

References

[1] James R B, Schlesinger T E, Siffert P and Franks L eds 1993 *Semiconductors for room-temperature radiation detector applications* vol 302 (Pittsburgh, PA: Materials Research Society)
[2] Furdyna J K 1988 *J. Appl. Phys.* **64** R29–R64
[3] Chen C-C and Whitaker J F 2008 *Appl. Phys. Lett.* **92** 101119
[4] Rey-de-Castro R, Wang D, Zheng X, Verevkin A, Sobolewski R, Mikulics M, Adam R, Kordoš P and Mycielski A 2004 *Appl. Phys. Lett.* **85** 3806–08
[5] Cross A S, Knauer J P, Mycielski A, Kochanowska D, Wiktowska-Baran M, Jakieła R, Domagała J, Cui Y, James R B and Sobolewski R 2010 *Nucl. Instrum. Methods Phys. Res. A* **624** 649–55
[6] Hossain A, Yakimovich V, Bolotnikov A E, Bolton K, Camarda G S, Cui Y, Franc J, Gul R, Kim K H, Pittman H, Yang G, Herpst R and James R B 2013 *J. Cryst. Growth* **379** 34–40
[7] Slagle J E, Lombardo D, Trivedi S and Guha S 2015 *Proc. SPIE* **9359** 93591W
[8] Mycielski A, Kowalczyk L, Gałazka R R, Sobolewski R, Wang D, Burger A, Sowińska M, Groza M, Siffert P, Szadkowski A, Witkowska B and Kaliszek W 2006 *J. Alloys Compd.* **423** 163–68
[9] Serafini J, Hossain A, James R B, Guziewicz M, Kruszka R, Słysz W, Kochanowska D, Domagala J Z, Mycielski A and Sobolewski R 2017 *Appl. Phys. Lett.* **111** 011108
[10] Ketchen M B, Grischkowsky D, Chen T C, Chi C-C, Duling III I N, Halas N J, Halbout J-H, Kash J A and Li G P 1986 *Appl. Phys. Lett.* **48** 751–53
[11] Yariv A 1997 *Optical electronics in modern communications* 5th ed. The oxford series in electrical and computer engineering (New York: Oxford University Press)
[12] Alexandrou S, Wang C-C, Currie M, Sobolewski R and Hsiang T Y 1994 *Proc. SPIE* **2149** 108–18
[13] Chenault D B, Chipman R A and Lu S-Y 1994 *Appl. Opt.* **33** 7382–89

IOP Conf. Series: Journal of Physics: Conf. Series **906** (2017) 012017 doi:10.1088/1742-6596/906/1/012017

Quantum walks of two interacting particles on percolation graphs

Ilaria Siloi[1], **Claudia Benedetti**[2], **Enrico Piccinini**[3], **Matteo G. A. Paris**[2], **Paolo Bordone**[4]

[1] Department of Physics, University of North Texas, Denton, TX 76203, USA.
[2] Quantum Technology Lab, Dipartimento di Fisica, Università degli Studi di Milano, I-20133, Milano, IT
[3] ARCES Research Center, Via Toffano 2/2, I-40125, Bologna, IT
[4] Dipartimento di Scienze Fisiche, Informatiche e Matematiche, Università degli Studi di Modena e Reggio Emilia, I-41125, Modena, IT

E-mail: paolo.bordone@unimore.it

Abstract. We address the dynamics of two indistinguishable interacting particles moving on a dynamical percolation graph, i.e., a graph where the edges are independent random telegraph processes whose values jump between 0 and 1, thus mimicking percolation. The interplay between the particle interaction strength, initial state and the percolation rate determine different dynamical regimes for the walkers. We show that, whenever the walkers are initially localised within the interaction range, fast noise enhances the particle spread compared to the noiseless case.

1. Introduction

Quantum walks (QWs) are the quantum analogue of classical random walks and describe the propagation of quantum particles over a discrete lattice with non zero tunneling amplitudes between adjacent sites [1]. Two-particles quantum walks are paradigmatic systems to address the interplay between particle indistinguishability and particle interaction. Besides the fundamental interest, two-particle quantum walks are implemented on different platforms also with the aim of studying multiple quantum interference and to simulate physical, chemical, and biological complex systems [2]. Experimental realisations of QWs are subject to different sources of noise e.g. imperfections, defects or external perturbations - that may dramatically affect the dynamical behaviour of the walkers [3, 4]. As a consequence of the external noise, the regular structures underlying the QWs may be altered, and the ordered lattices may turn into irregular graphs. Among dynamically varying graphs, a central role is played by dynamical percolation graphs, which are random graphs where edges are created and destroyed in time, according to some stochastic process. While, to analyse more realistic scenarios, in a previous work [4] we focused on the case of a non-Gaussian random telegraph noise which randomises the tunneling amplitudes between adjacent sites still retaining for them a finite value, in this paper we address the decoherent dynamics of two indistinguishable and interacting particles over one-dimensional percolation graphs, with links that appear and disappear randomly in time. The percolation approach is indeed quite intriguing in quantum mechanics, and can even be thought as related to the cellular automaton interpretation of quantum mechanics itself [5].

2. The physical model

The continuous-time quantum walk of two indistinguishable particles on percolation graphs is described by the Hamiltonian:

$$H_2(t) = H_1(t) \otimes \mathbb{I} + \mathbb{I} \otimes H_1(t) + H_{\text{int}} \tag{1}$$

where the single-particle Hamiltonian H_1 and the interaction Hamiltonian H_{int} read:

$$H_1 = \epsilon\mathbb{I} + \sum_x J g_x(t)\Big(|x\rangle\langle x+1| + |x+1\rangle\langle x|\Big), \quad H_{\text{int}} = U_p(|x-y|)\sum_{x,y=1}^{N} |x,y\rangle\langle x,y|. \tag{2}$$

$\mathbb{I}$ is the single-particle identity matrix and the term $\epsilon\mathbb{I}$ only determines a rescaling of the energies and can be omitted without loss of generality, J is the coupling constant, $\{|x\rangle\}$ represents the orthonormal basis of the site of the graphs, i.e. the walkers positions, while the set $\{|x,y\rangle\}$ describes the orthonormal basis for the two-particle QW, where one occupies site x and the other site y. $g_x(t)$ is a stochastic process describing the percolating links, i.e. it is described as the non-Gaussian random telegraph noise with a switching rate γ and autocorrelation function $C(t) = \langle g_x(t)g_y(0)\rangle = \delta_{x,y}\frac{1}{4}e^{-2\gamma t}$, whose values jump between 0 and 1, indicating the absence and presence of edges respectively [3]. U_p is the strength of the interaction between the two walkers, as a function of their distance:

$$U_p(|x-y|) = \begin{cases} U & \text{if} \quad x = y \text{ and } p = \text{bosons} \\ U/3 & \text{if } |x-y| = 1 \text{ and } p = \text{fermions} \end{cases}, \tag{3}$$

i.e. bosons experience on-site interaction energy while fermions only nearest-neighbours interaction. The initial state of the two walkers is taken in the form: $|\psi_0\rangle = \frac{1}{\sqrt{2}}(|x,y\rangle \pm |y,x\rangle)$ to take into account their indistinguishability. Since Eq. (1) conserves the symmetry of the wavefunction, the initial state of the walkers fixes the statistics of the particles. The time evolution of the initial state is determined by ensemble-averaging the dynamics over all possible realisations of the noise $\langle \Lambda_r \rho_0 \Lambda_r^\dagger(t)\rangle_{\{g_x(t)\}}$ where $\Lambda_r(t) = \mathcal{T}e^{-i\int_0^t H_2(s)ds}$ is the evolution operator for the single realisation , $\mathcal{T}$ is the time-ordering operator and $\rho_0 = |\psi_0\rangle\langle\psi_0|$. Hereafter, we use natural unit $\hbar = 1$ and we set $J = 1$ such that all quantities will be given in unit of the coupling constant.

3. Results and discussion

In order to characterise the dynamics of the two-particle QW on percolation graphs, we consider a 1D lattice with N=80 nodes, and we fix the switching rate of the percolating links $\gamma = 10$ (in unit of J). We simulate the dynamics, by numerically calculating the ensemble average over 1000 different percolation realisations, using a specific GPU accelerated code [7]. In order to characterize the dynamics, we evaluate the temporal behaviour of the single particle variance $\sigma^2(t) = \sum_x \langle x^2(t)\rangle - \langle x(t)\rangle^2$, with $\langle x^k\rangle = \sum_i i^k \rho^1_{ii}(t)$ and ρ^1 the single particle density matrix. To complete our analysis, we also evaluate the occupation number of the lattice sites during the evolution $\langle n_k(t)\rangle = 2\sum_j \rho_{kj,kj}(t)$, where $\rho_{kj,kj}(t)$ are the populations of the two-particle density matrix in the $\{|k,j\rangle\}$ basis. In the following we present our results for both bosons and fermions.

In Fig. 1 we show the behaviour of the single-particle variance as a function of time (in unit of J), for both bosons and fermions. In each panel we compare the dynamics for the regular lattice without noise (black lines) and the case of percolation graphs (dashed red lines). In the left part of the figure, we consider nearest-neighbours sites as initial state for fermions, while bosons initially occupy the same node. An important result emerges here: for suitable interaction strength regime ($U < 6$ for fermions and $U < 2$ for bosons), the single-particle variance for

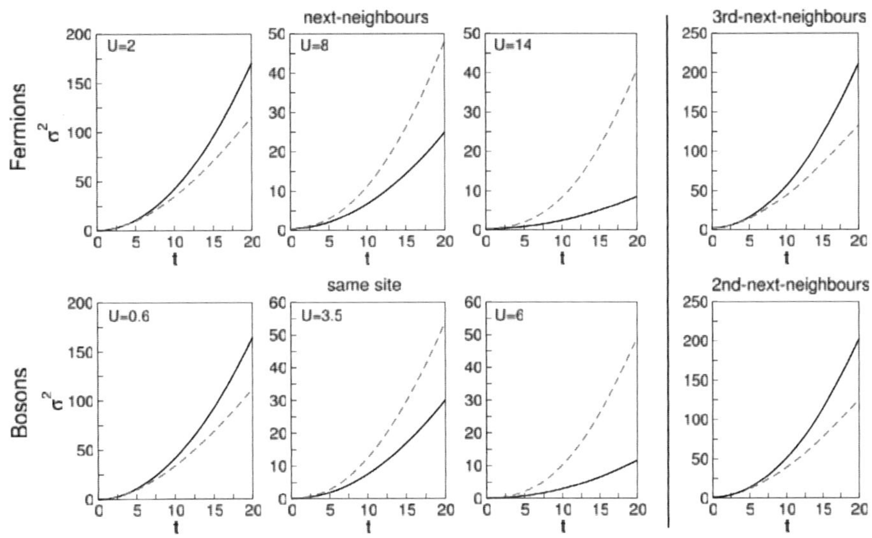

Figure 1. Single-particle variance as a function of dimensionless time t over a graph of $N = 80$ sites, for different values of the interaction strength U. Solid black lines are for regular noiseless lattices, while the dashed red lines represent the case of percolation graphs with $\gamma = 10$. The two panels in the right part of the figure are for fermions and bosons starting in third-next and second-next-neighbours respectively. In the latter case the results are essentially independent on the value of U.

percolation graphs is smaller with respect to the noiseless case, independently on the particle statistics. But as the interaction strength grows, we observe a crossover in its behaviour, and σ^2 for percolation graphs exceeds the noiseless case. This means that the particles can spread faster over sites in the lattice, due to fast ($\gamma = 10$) percolation. This happens when the two particles are initially located on nearest-neighbours sites (fermions) or same-site (bosons), that is, if they are positioned within the range of interaction (3). One may notice that particles undergo a different dynamics if the initial positions are outside the range of interaction (U=0), as shown in the rightmost panels of Fig. 1 where the two particles start from third-next (fermions) and second-next (bosons) -neighbours. In this case, indeed, the crossover between percolation and noiseless variance never happen and the free walkers always spread faster on the noiseless graph with respect to the percolation graph.

The effect of dynamical percolation is also further analysed by studying the occupation number $\langle n(t) \rangle$ of the lattice sites. This is shown in Fig. 2. The top panels show the distribution of the occupation number in the noiseless case, for strong interactions. As known [4], the particles tend to localise, with minor components of the wavefunction travelling away from the initial sites. The situation changes if we consider the dynamics of the QW over a percolation graph. Indeed, the particles spread more over the lattice, with wavefunction components that occupy a significantly larger number of the nodes, in agreement with the behaviour of the single-particle variance. This dynamical behaviour can be explained in terms of the band structure of the Hubbard model. In the absence of interaction (U=0), the band structure consist of a unique band identical for bosons and fermions. For finite interaction strength, a small band is formed, called mini-band, whose energy at the edge of the first Brillouin zone is given approximately by $U/3$ for fermions and U for bosons. The remaining states are contained in the main band, which ranges approximately from -4J to 4J [6]. Even if it is not possible to define proper band structures since the translational invariance is broken by the noise, we can consider fast percolating noise

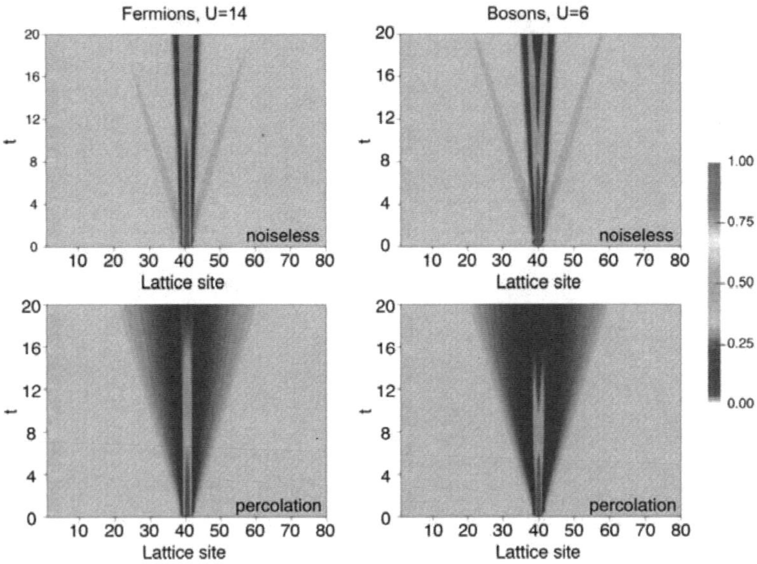

Figure 2. Occupation number $\langle n_k(t) \rangle$ as a function of sites k and dimensionless time t, for fermions and bosons in the noiseless and percolation lattice of $N = 80$ sites. Particle interaction is set $U = 14$ for fermions and $U = 6$ for bosons. Fermions start in nearest-neighbours sites, while bosons are initially localised on the same site.

as a perturbation to the noiseless dynamics and, to first-order approximation, we can explain the QW features in terms of bands [4]. In this framework, we can state that noise provides a redistribution of the wavefunction components among the subbands and it allows to access a new regime where the particles acquire faster propagation contributions, thus breaking the localisation induced by the strong interaction. This redistribution and the consequent crossover of the single-particle variance happens whenever the particles initially belong to the mini-bands and, because of noise, they are allowed to access other sub-bands. On the contrary, whenever particles are located outside the range of interaction, i.e. in the main sub-band, noise cannot add any faster contribution, thus the noiseless dynamics is always faster than the one obtained on a percolation graph.

For the sake of completeness, we comment also on the dynamics of two-particle QW in the presence of slow percolation, i.e. when $\gamma \ll 1$. The percolation rate in this case is very small and in the limit we reach a static percolation regime. This situation does not allow for any propagation of the walker and the combined effect of noise and interaction strongly localise the particle on their initial sites.

We conclude by remarking that localisation-breaking noise is an important feature that can be exploited in many contexts especially for protocols for quantum transport and communication, where noise can be engineered in order to obtain an enhancement of quantum properties.

This work has been supported by EU (QuProCS Grant Agreement No. 641277), by NVIDIA (Titan X Pascal GPU) and by UNIMORE through FAR2014. The authors thank S. Maniscalco and J. Piilo for fruitful discussions.

References

[1] Kempe J 2003 *Cont. Phys.* **44** 307.
[2] Sansoni L *et al.* 2012 *Phys. Rev. Lett.* **108** 010502; Mohseni M *et al.* 2008 *J. Chem. Phys.* **129** 174106; Preiss P M *et al.* 2015 *Science* **347** 1229.
[3] Benedetti C, Buscemi F, Bordone P, Paris M G A 2016 *Phys. Rev* A **93** 042313.
[4] Siloi I, Benedetti C, Piccinini E, Piilo J, Maniscalco S, Paris M G A, and Bordone P 2017 *Phys. Rev. A* **95** 022106.
[5] 't Hooft G 2016 *The Cellular Automaton Interpretation of Quantum Mechanics* (Springer).
[6] Scott A, Eilbeck J, Gilhøj H 1994 *Physica D* **78** 194; Valiente M and Petrosyan D 2008 *J. Phys. B* **41** 161002.
[7] Piccinini E, Benedetti C, Siloi I, Paris M G A and Bordone P 2017 *Compt. Phys. Comm.* **215** 235.

IOP Conf. Series: Journal of Physics: Conf. Series **906** (2017) 012018 doi:10.1088/1742-6596/906/1/012018

Drift and diffusion high-field magneto-transport in GaN

G. I. Syngayivska[1], V. V. Korotyeyev[1], V. A. Kochelap[1] and L. Varani[2]

[1]Department of Theoretical Physics, Institute of Semiconductor Physics, Kyiv 03028, Ukraine
[2]Institut d'Electronique du Sud, CNRS UMR 5214, University Montpellier 2, France

E-mail: `koroteev@ukr.net` , `singg@ukr.net`

Abstract. The paper is focused on the investigation of magneto-transport phenomena in the compensated bulk-like GaN sample. Particularly, we studied the diffusion coefficient of the electrons in parallel and crossed configurations of moderate electric (E=1...10 kV/cm) and magnetic (H=1...4 T) fields. We found that E-field dependencies of the transverse-to-current diffusion coefficient are non-monotonic for both configurations with magnitude of the diffusion coefficient greatly controlled by the H-field. We showed that different behavior of the diffusion processes corresponds to distinct kinetics of the hot electrons. We suggest that measurements of the diffusion coefficient under E- and H-fields will allow to identify important for applications regimes of the electron kinetics.

1. Introduction

Wide bandgap semiconductor compounds, such as GaN, InN, AlN and quantum heterostructures on their basis have great perspectives for various applications in the modern high-power and high-frequency micro- and optoelectronics [1]. The unique properties of the nitrides – high low-field mobility, strong electron-optical-phonon coupling and large optical phonon energy ($\hbar\omega_0 \approx 90$ meV for GaN) is favorable for the realization of streaming electron transport that associated with quasi-ballistic electron motion in momentum space that leads to the formation of strongly anisotropic distribution function [2, 3]. This effect is possible at low temperatures, $k_B T_0 < \bar{h}\omega$ and small electron concentrations, where e-e scattering does not prevail and the electrons have independent *individual* momentum and energy budgets. Previously, it was shown for GaN structures that streaming effect can occur at electron concentration of $< 10^{16}$ cm^{-3} (for bulk-like samples)[4] and $< 5 \times 10^{11}$ cm^{-2} (for quantum heterostructures with 2D electron gas)[5] at temperatures < 150 K in the range of applied electric fields of $3 - 10$ kV/cm. Formation of the streaming-like electron kinetics is characterized by the emergence of the specific saturation behavior of current-voltage characteristics, anisotropy of electron conductivity and frequency ranges of negative dynamic conductivity (the effect of optical phonon transit-time (OPTT) resonance)[4, 5, 6]. OPTT resonance can be used as core mechanism for a development of electrically-pumped generators in THz frequency range. Additionally, the useful information on formation of streaming transport regime can be obtained by investigation of magnetic field effects. For example, in parallel configuration of applied electric and magnetic fields ($\vec{E}||\vec{H}$) it can be observed co-existence of OPTT and cyclotron resonances. Latter can be identified by strong dependencies of transmission/absorption spectra vs polarization of incident radiation[7, 8].

If the magnetic field is oriented perpendicular to the electric field, the electron dynamics can be essentially changed. Particularly, the magnetic field can destroy the streaming transport forming a vortex-like electron motion which suppresses the main energy dissipation due to optical phonon emission[9]. It can leads to the formation of population inversion and a series of galvanomagnetic effects such as a negative magneto resistance and collapse of the dissipative current [10, 11]. The aim of the paper is the study of the magnetic field effect on diffusion processes of hot electron in coordinate space.

2. Diffusion coefficients in $\vec{E}||\vec{H}$ configuration

Diffusion coefficient is calculated using Monte Carlo simulation [12] of semiclassical magneto–transport in bulk-like GaN sample. In contrast of previous studies of diffusion coefficients in nitrides [13] we restrict ourself by the case of moderate applied electric fields when the effect of inter-valley transitions is negligibly small. To exclude the quenching effect on the streaming regime by *e-e* scattering, in the transport model the sample is supposed to be compensated, the electron concentration n_e is less than impurity one, N_i. In the simulation we take into account the three main scattering mechanisms: scattering by ionized impurities, acoustic phonons and polar optical phonons. The details of transport model is given in refs.[7, 11]. The magnetic field is treated as a classical one.

In this section the geometry of the applied fields is following: both $\vec{E}$ and $\vec{H}$ are oriented along $z-$axis. Early, we found that at magnetic fields of a moderate strength ($H < 10$ T), the steady-state characteristics such as the distribution function of hot-electrons, the drift velocity, V_d (E), and the average energy, $< \epsilon >$ (E) practically are independent of the magnetic field [7]. This is a consequence of two main factors: (i) for the parabolic electron dispersion and the parallel configuration of electric and magnetic fields, electron motion along fields direction and that in perpendicular plane are uncoupled; (ii) the magnetic field does not affect the energy balance of the electron system. However, the electron diffusion process in coordinate space demonstrates the essential dependence vs values of H. Electric field dependences of transversal component of the diffusion coefficient, $D_{xx}(E)$, calculated at three given magnitudes of H are shown in Fig.1(a) by solid lines. As seen, the dependence of $D_{xx}(E)$ has non-monotonic

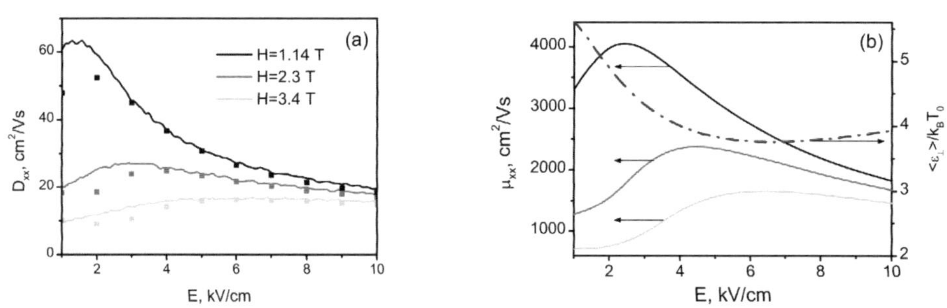

Figure 1. Panel (a): dependences of $D_{xx}(E)$ (solid lines) and $D_{xx}^{(a)}(E)$ (dots). Panel (b): dependences of $\mu_{xx}(E)$ (solid lines) and $\epsilon_\perp(E)$ (dash-dotted line). Calculations performed at $N_i = 10^{16}$ cm^{-3}, $n_e = 10^{15}$ cm^{-3} and $T_0 = 30$ K.

behavior and electron diffusion in the perpendicular direction to the applied fields is essentially suppressed with increasing of H. Such behavior of $D_{xx}(E)$ is correlated with field dependences of transversal component of low-field mobility μ_{xx} (shown in Fig.1(b) by solid lines). We found that field dependencies of $D_{xx}(E)$ can be approximately described by the phenomenological Einstein relationship, $D_{xx}^{(a)} =< \epsilon_\perp > \mu_{xx}/e$, where $< \epsilon_\perp >$ is the part of the average electron energy corresponding to the electron motion in perpendicular plane. In the frames of the assumed transport model this quantity is the independent on H and decreases in the range of electric fields corresponding to the formation of developed streaming transport regime (see dashed-dotted line in Fig.1(b)). The results of the diffusion coefficient approximation by the Einstein relationship (dots in Fig. 1(a)) are in the good agreement with exact calculations.

3. Diffusion coefficients in $\vec{E}\perp\vec{H}$ configuration

In this section we assume that $\vec{E}$ and $\vec{H}$ are oriented along $z-$axis and $y-$axis, respectively. In such geometry of the crossed electric and magnetic fields the electron kinetics is much more complicated. The presence of magnetic field can essentially affect on the form of both steady-state distribution function and transport characteristics[11]. Depending on the ratio between magnitudes of the electric and magnetic fields, it can be realized three typical situations. In the case (I) of the weak magnetic field ($H < H_1 \equiv m^* cE/P_0$, where $P_0 = \sqrt{2m^*\hbar\omega_0}$, c is the light velocity) electron trajectories in the momentum space are weakly bent and the streaming motion is maintained. The form of distribution function remains strongly-anisotropic and elongated along $z-$axis. In the case (II) of the intermediate magnetic fields ($H_1 < H < 2H_1$) there are two groups of electrons. In the former group, the electrons execute a streaming cyclic motion over bent trajectories. In the latter group, the electrons rotate over the Larmor orbits and do not interact with optical phonons. The effect of accumulation of the essential part of electrons on the Larmor orbits stipulates the formation of the complicated topological structure of distribution function. In the case (III) of strong magnetic fields ($H > 2H_1$) most of the electrons are captured on the Larmor orbits. As a result the optical phonon emission is strongly suppressed, distribution function becomes quasi-isotropic and acquires the axial symmetry. This leads to the decreasing of the Ohmic and Hall components of current [11].

We analysis the field dependencies of diffusion coefficient components in transversal directions to $\vec{E}$. Electric field dependences of $D_{xx}(E)$ are shown in Fig.2(a). As seen, the increasing of magnetic field is considerably suppressed electron diffusion in the $x-$ direction. The quantity $D_{xx}(E)$ at finite H as well at $H = 0$ exhibits a non-monotonic behavior with the maximum. At higher magnetic fields the maximum shifts to the region of higher electric fields. The positions of the maxima are marked in Fig.2(b) by circles. All circles lie in the region of $E-H$ corresponding to the case (II) of the co-existence of two electron groups.

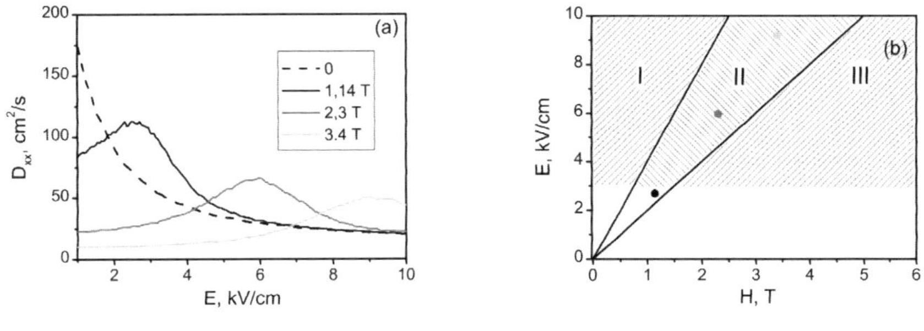

Figure 2. Panel (a): dependences of $D_{xx}(E)$. Other parameters are the same as in Fig.1. Panel (b): Diagram of $E - H$ regions corresponding to the (I), (II), (III) cases of electron kinetics in crossed fields. Shadow demarcates the electric fields of developed streaming regime.

The field dependencies of the D_{yy} component which describes the electron diffusion along H cardinally differ from the field dependencies of the D_{xx} component. As seen in Fig.3(a) the magnetic field stimulates the growth of electron diffusion along magnetic field. The obtained values of the D_{yy} are in several times larger than values of the D_{yy} at the same H. It should be noted that magnetic field weakly affect on the position of maximum observed in the $D_{yy}(E)$ dependencies. At such high electric fields when the case (I) occurs the $D_{yy}(E)$ tends to the constant value which is independent on both E and H. We found that general behavior of D_{yy} component of diffusion coefficient is correlated with electric field dependence of quantity $< \epsilon_y >$ that is the part of the average electron energy corresponding to the motion along $y-$direction (see Fig.3(b)).

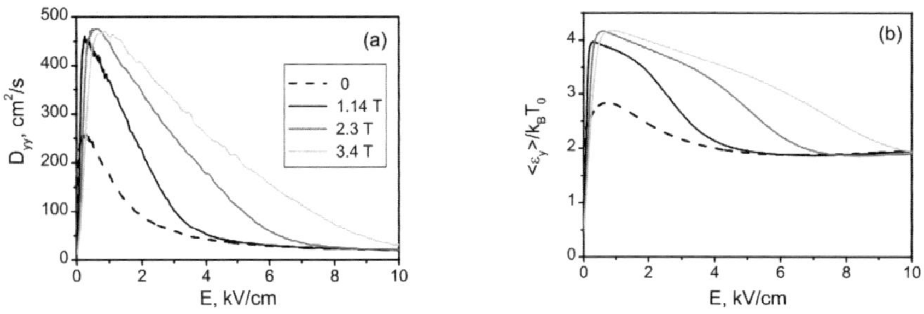

Figure 3. Panel (a): dependences of $D_{yy}(E)$. Panel (b): dependences of $< \epsilon_y > (E)$. Other parameters are the same as in Fig.1.

In summary, investigation of the diffusion processes in electric and magnetic fields can provides valuable information on kinetic regimes of hot electrons in GaN. Besides, study of the field dependencies of the diffusion coefficient are important for many applications, device modeling, characterization of noise and space-charge phenomena in hot-electron devices, etc. Also, such a study can be useful for development of novel types of magnetic field sensors, which exploit the specific features of ballistic and quasi-ballistic transport regimes[14, 15].

Acknowledgments

Paper was supported by BMBF Project EUV-TERA with funding number 01DK17028.

References

[1] Millan J, Godignon P, Perpina X, Perez-Tomas A and Rebollo J 2014 *IEEE Transactions on Power Electronics* **29**(5) 2155-63
[2] Korotyeyev V V, Kochelap V A, Kim K W and Woolard D L 2003 *Appl. Phys. Lett.* **82** 2643
[3] Starikov E, Shiktorov P, Gruzinskis V, Varani L, Palermo C, Millithaler J-F and Regiani L 2007 *Phys. Rev.* B **76** 045333
[4] Syngayivska G I and Korotyeyev V V 2013 *Ukr. J. Phys.* **58** 40-55
[5] Lu J T and Cao J C 2005 *Semicond. Sci. Technol.* **20** 829
[6] Starikov E, Shiktorov P, Gruzinskis V, Varani L, Palermo C, Millithaler J-F and Regiani L 2008 *J. Phys.: Condens. Matter* **20** 384209
[7] Syngayivska G I, Korotyeyev V V and Kochelap V A 2013 *Semicond. Sci. Technol.* **28** 035007
[8] Korotyeyev V V 2013 *Semiconductor Physics, Quantum Electronics & Optoelectronics* **16**(1) 18
[9] Vosilius I I and Levinson I B 1966 *JETP* **23** 1104; 1967 *JETP* **25** 672
[10] Komiyama S 1982 *Phys. Rev. Lett.* **48** 271
[11] Syngayivska G I, Korotyeyev V V, Kochelap V A and Varani L 2016 *J. Appl. Phys.* **120** 095704
[12] Fauquembergue R, Zimmermann J, Kaszynski A, Constant E and Microondes G 1980 *J. Appl. Phys.* **51**(2) 1065
[13] Starikov E, Shiktorov P, Gruzinskis V, Reggiani L, Varani L, Vaissiere J C and Palermo C 2005 *Semicond. Sci. Technol.* **20** 279
[14] Gilbertson A M, Benstock D, Fearn M, Kormanyos A, Ladak S, Emeny M T, Lambert C J, Ashley T, Solin S A and Cohen L F 2011 *Appl. Phys. Lett.* **98** 062106
[15] Santoruvo G, Allain A, Ovchinnikov D and Matioli E 2016 *Appl. Phys. Lett.* **109** 103102

The issue of 0D-like ground state isolation in GaAs- and InP-based coupled quantum dots-quantum well systems

M. Syperek[1], J. Andrzejewski[1], W. Rudno-Rudziński[1], A. Maryński[1], G. Sęk[1], J. Misiewicz[1], J. P. Reithmaier[2,3], A. Somers[2], S. Höfling[3,4]

[1] Laboratory for Optical Spectroscopy of Nanostructures, Department of Experimental Physics, Faculty of Fundamental Problems of Technology, Wrocław University of Science and Technology, Wybrzeże Wyspiańskiego 27 27, 50-370 Wrocaw, Poland
[2] Institute of Nanostructure Technologies and Analytics (INA), CINSaT, University of Kassel, Heinrich-Plett-Str. 40, 34132 Kassel, Germany
[3] Technische Physik, University of Würzburg and Wilhelm-Conrad-Röntgen-Research Center for Complex Material Systems, Am Hubland, D-97074 Würzburg, Germany
[4] SUPA, School of Physics and Astronomy, University of St. Andrews, North Haugh, KY16 9SS St. Andrews, United Kingdom

E-mail: marcin.syperek@pwr.edu.pl

Abstract. The issue of quantum mechanical coupling between a semiconductor quantum dot and a quantum well is studied in two families of GaAs- and InP- based structures at cryogenic temperatures. It is shown that by tuning the quantum well parameters one can strongly disturb the 0D-character of the coupled system ground state, initially located in a dot. The out-coupling of either an electron or a hole state from the quantum dot confining potential is viewed by a significant elongation of the photoluminescence decay time constant. Band structure calculations show that in the GaAs-based coupled system at its ground state a hole remains isolated in the dot, whereas an electron gets delocalized towards the quantum well. The opposite picture is built for the ground state of a coupled system based on InP.

1. Introduction

The coupling between semiconductor self-assembled quantum dots (QDs) and a quantum well (QW) is considered as an essential process that would bring new functionalities to electronic or optoelectronic devices. In the concept, a QW serves as a carrier mobility channel that can mediate the charge or spin exchange process between a QD state ultimately defined in the real space. The coupling is a purerly quantum mechanical effect where QDs are separated from a QW by a thin enough barrier, allowing charge or spin tunneling between subsystems. Although, the idea is simple in its form, however, to achieve a certain functionality one need carefully guide the engineering process of the coupled QDs-QW (CQDsQW) system to avoid undesirable consequences of the coupling. A striking example is the necessity to preserve an isolated 0D-like character of the QD state in the environment determined by the electronic coupling. Such state is of primary importance for many applications based on the CQDsQW concept. One can only mention tunnel injection lasers[1, 2] or spin memory devices[3]. In the first example, the coupling results in an efficient carrier redistribution process among a dense ensemble of QDs but to get benefits from the QDs as a laser gain medium one need assure the 0D-like density of states for the lasing state. In the case of a spin memory device, the QW plays a role of the information highway that can address a QD memory element. However, the spin state must be efficiently isolated from other states acting as a source of decoherence.

This work reveals the issue of losing the isolated character of the QD-like ground state (GS) in the CQDsQW system that can be observed in two technologically important material systems based on GaAs and InP. The work is experimental, with the leading role of time-resolved photoluminescence spectroscopy (TRPL), supported by photoluminescence (PL), and modulated reflectivity (ΔR/R), but also theoretical calculations of the CQDsQW band structure.

2. Experimental details and methodology

Investigated CQDsQW structures were grown by molecular beam epitaxy either on a semi-insulating (001)GaAs or a sulfur-doped (001)InP substrate.

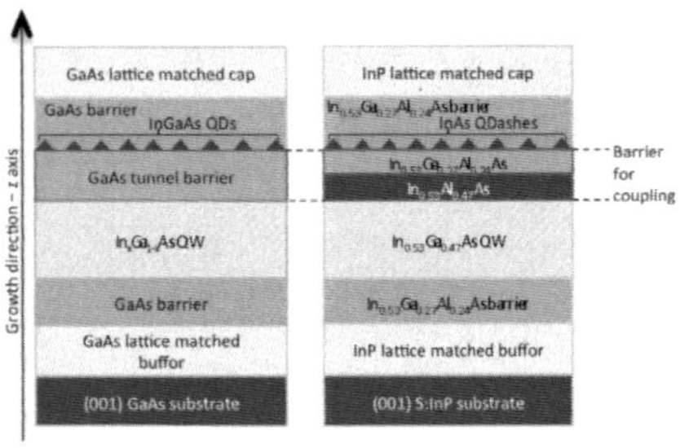

Figure 1. The layer sequence of coupled QDs-QW structures based on GaAs (left hand side) and InP substrate (right hand side).

The layer sequence for considered systems is depicted in Fig. 1 and the growth details are presented in Ref.[4, 5]. QDs are different in both material systems. The (In,Ga)As/GaAs QDs are quite typical, with a pyramidal shape of 16 nm in the base length and 1.7 nm in height. For such dots, the GS energy is centered at ~1.22 eV. The InAs/InGaAlAs/InP QDs are strongly elongated in one of the in-plain direction, having a triangular cross section, with the base width of ~30 nm and height of ~3.5 nm.[6] The typical length of the dot exceeds 150 nm, which is why these structures are sometimes called quantum dashes (QDashes). The ground state in these structures is centered near 0.85 eV, right at the third telecom window. The coupling between the QW and QD GSs is tuned by adjusting parameters of the well, while keeping the barrier width constant (2 nm). With the QW parameters of d=15 nm (well width), and the [In] content x=0.17, the GaAs-based CQDsQW GS is nominally uncoupled, as compared to the coupled structure with d=7 nm and x=0.23. For the InP-based structure, the nominally uncoupled CQDsQW GS is obtained for d=4.5 nm as compared to the coupled one, with d=6.5 nm. In both cases, the chemical content of the well remained unchanged.

For the continuous-wave (CW) PL, ΔR/R, and TRPL experiments, the structures were held at T=4.2 K in a He-flow cryostat. The 532 nm line from the YAG laser was used as a source of photo-excitation in the CW regime. For the TRPL experiment, structures were excited by a train of 140-fs long pulses, at a repetition frequency of 76 MHz generated by the Ti:Sapphire oscillator. Subsequent PL emission was dispersed by a 0.3-m-focal-length monochromator and detected by a streak camera system equipped either with the S1 or liquid nitrogen-cooled NIR photocathode. The CW PL was measured in the same setup, with the InGaAs linear array detector. Details of the PR experiment are described in Ref.[6, 7].

Calculations of the CQDsQW system band structure was performed in the 3D eight-band **kp** model[4, 8], whose implementation includes strain fields and piezoelectric effects.

IOP Conf. Series: Journal of Physics: Conf. Series **906** (2017) 012019 doi:10.1088/1742-6596/906/1/012019

3. Results and discussion

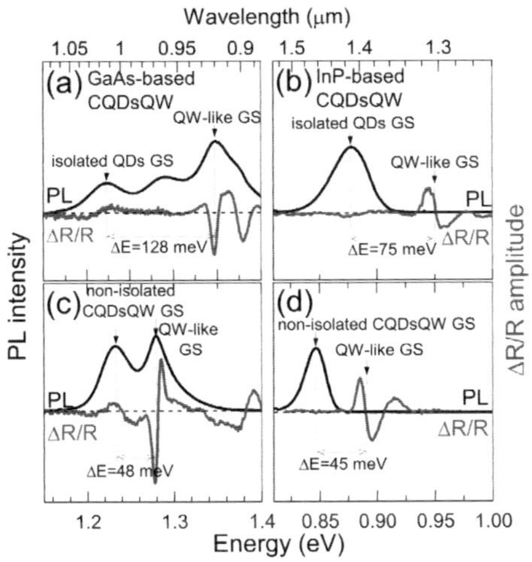

Figure 2. (a) PL and ΔR/R spectra for the GaAs-based CQDsQW system with the uncoupled ground state isolated within a QD, and (b) its InP-based counterpart. (c) PL ΔR/R spectra for the GaAs-based CQDsQW system with the coupled GS, and (d) its InP-based counterpart. T=4.2 K

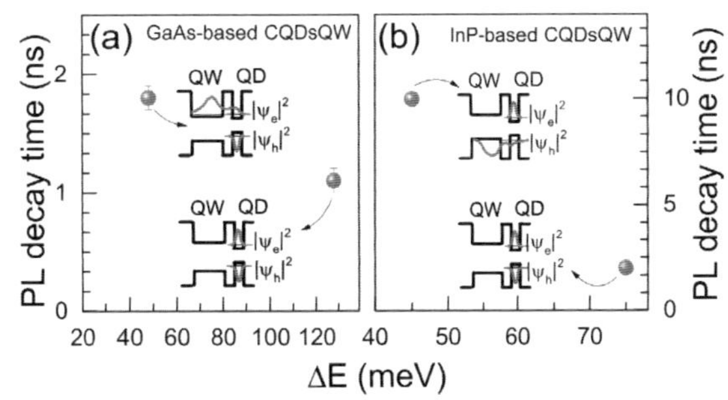

Figure 3. PL decay time obtained for GS emission of (a) GaAs-based CQDsQW system, and (b) InP-based one. T=4.2 K and non-resonant excitation at 830 nm. The idea concerning the lack or presence of isolation of the CQDsQW system GS is depicted in sketches. $|\Psi_e|^2$ and $|\Psi_h|^2$ are single particle, electron and hole density probability functions.

The PL spectra for CQDsQW structures characterized by the uncoupled and coupled GS are depicted in Fig. 2 (black line). Since the coupling at the GS is hard to parametrize, one can use the ΔE value that defines the energy distance between the CQDsQW GS and the first QW-like GS transition revealed as a well visible absorption-like feature in the ΔR/R spectrum in Fig.2(violet line). The significant value of ΔE may indicate that the presence of the QW confining potential disturbs only the higher lying states of a QD, while the coupled system GS remains isolated in a dot. The TRPL experiment confirms the hypothesis. The PL lifetime measured at the GaAs-based CQDsQW GS is ~1 ns and ~2 ns for the InP-based structure with a large ΔE. Both numbers are similar to those presented for typical QDs/QDashes for a material system considered here. By modifying the QW parameters for specified application purposes, one can obtain a much deeper confining potential of the well. Its presence may obviously disturb the lowest lying states of a QD. It can happen for the structures with small ΔE presented in Fig. 2 (c) and (d). In this case, the lack of isolation of a QD GS may result in the out-coupling of the particle probability density function from the confining potential of the dot. This, however, would be translated into an elongated PL decay time measured at the CQDsQW GS. Indeed, the PL decay time increases from ~1 ns to ~1.9 ns for the GaAs-based structure when such out-coupling is expected. A similar behavior is observed for the InP-based counterpart, where the decay time increases from ~2 ns to ~10 ns. Although experimental evidence for the lack of 0D-like isolation at the coupled system GS seems to be clear for specific QW parameters, there is still a question concerning the outcoupling nature in both investigated material systems. To address this issue, one can perform the band structure calculations of CQDsQW system.

Figure 4 shows potential band profiles for GaAs- and InP-based CQDsQW systems in the absence (Fig. 4(a),(c)) and presence of the GS coupling (Fig. 4(b),(d)). One can see that in the

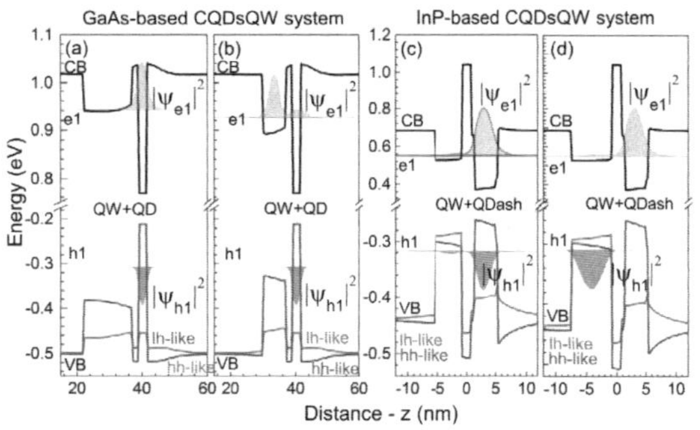

Figure 4. Calculated band structure for GaAs- and InP-based CQDsQW with QW parameters (a) d=15 nm, In content: x=0.17, (b) d=7 nm, x=0.23, (c) d=4.5 nm, (d) d=6.5 nm. $|\Psi_{e1,h1}|^2$ are single particle, electron or hole, probability density functions, CB and VB are conduction and valence bands for two types of holes (hh-like, lh-like).

case of the GaAs-based system, a hole state is isolated in the dot, whereas an electron one is smeared over the QD and QW potential. Surprisingly, the opposite picture appears for the InP-based coupled system. In this case an electron state is effectively localized within the dot/dash structures, whereas a hole is delocalized among the effective potential created by the dot/dash and the QW.

4. Conclusions

The required coupling between a semiconductor QD and a QW makes a design process of the coupled QD-QW system quite challenging. Tuning the QW parameters towards achieving certain functionality of a device may lead to the loss of the 0D-like character of the coupled system GS, initially restricted to a QD. Moreover, depending on the material system, the nature of the out-coupling of the fundamental state can have a different origin. While in the GaAs-based coupled system there is a high chance to out-couple the electron state, keeping the hole one localized, in the InP-based systems the effect is exactly opposite. These results show that the design process of devices based on CQDsQW architecture is very demanding and needs a closer look, especially at material parameters, such as band offsets, to obtain expected benefits.

Acknowledgments

The work has been supported by the grant No. 2013/10/M/ST3/00636 of the National Science Centre in Poland and the QuCoS Project of Deutsche Forschungsgemeinschaft No. RE1110/16-1. The authors acknowledge technical support from Hamamatsu Co. with the NIR streak camera.

References

[1] G. Walter, N. Holonyak Jr., J. H. Ryou, and R. D. Dupuis 2001 *Appl.Phys. Lett.* **79** 1956
[2] Z. Mi, S. P. Bhattacharya, and S. Fathpour 2005 *Appl. Phys. Lett.* **86** 153109
[3] A. Marent, T. Nowozin, M. Geller, and D. Bimberg 2011 *Semicond. Sci.Technol.* **26** 014026.
[4] M. Syperek, J. Andrzejewski, W. Rudno-Rudziński, G. Sęk, J. Misiewicz, E. M. Pavelescu, C. Gilfert, and J. P. Reithmaier 2012 *Phys. Rev.* B **85** 125311
[5] W. Rudno-Rudziński, M. Syperek, J. Andrzejewski, A. Maryński, J. Misiewicz, A. Somers, S. Höfling, J. P. Reithmaier, and G. Sęk 2017 *AIP Adv.* **7** 015117
[6] W. Rudno-Rudziński, R. Kudrawiec, P. Podemski, G. Sęk, J. Misiewicz, A. Somers, R. Schwertberger, J. P. Reithmaier, and A. Forchel 2006 *Appl. Phys. Lett* **89**, 031908
[7] J. Misiewicz, P. Sitarek, G. Sęk, and R. Kudrawiec 2003 *Mater. Sci.* **21** 263
[8] J. Andrzejewski, G. Sęk, E. OReilly, A. Fiore, and J. Misiewicz 2010 *J. Appl. Phys.* **107** 073509

IOP Conf. Series: Journal of Physics: Conf. Series **906** (2017) 012020 doi:10.1088/1742-6596/906/1/012020

Screening effects of photogenerated carriers on terahertz radiation from coherent GaAs-like longitudinal optical phonons in (11*n*)-oriented GaAs/$In_{0.1}Al_{0.9}As$ strained multiple quantum wells

Hideo Takeuchi, Souta Asai, and Masaaki Nakayama

Department of Applied Physic, Graduate School of Engineering, Osaka City University, 3-3-138 Sugimoto, Sumiyoshi, Osaka 558-8585, Japan

Corresponding author's e-mail address: takeuchi@a-phys.eng.osaka-cu.ac.jp

Abstract. We investigated terahertz radiation from coherent GaAs-like longitudinal optical (LO) phonons in (11*n*)-oriented GaAs/$In_{0.1}Al_{0.9}As$ strained multiple quantum wells for clarifying the screening effects of photogenerated carriers. We observed the intense quasi-monochromatic terahertz wave from the coherent GaAs-like LO phonon, which originates from the initial polarization enhanced by the strong piezoelectric field. The intensity of the coherent GaAs-like LO-phonon band exhibited a saturation behavior as the pump power was increased. We evaluated the saturation behavior in terms of excitation efficiency of the terahertz wave from the coherent GaAs-like LO phonon using the parameter of unit-power intensity. From the pump-power dependence of the unit-power intensity, we conclude that the screening effect of high density photogenerated carriers on the piezoelectric field causes saturation of the terahertz-wave intensity from the coherent GaAs-like LO phonon.

1. Introduction

Femtosecond pulse-laser technology enables us to directly observe time-domain characteristics of phonons in phase referred as coherent phonons [1]. In polar semiconductors, the carriers generated by laser pulses release the initial amplitude of optical phonons through their instantaneous screening effects on internal electric fields. This starts the oscillation of the coherent optical phonons [2, 3]. There are two major techniques for observing the coherent phonons: pump-probe [1] and terahertz time-domain techniques [1, 4]. The initial amplitude of the coherent longitudinal optical (LO) phonon corresponds to the initial longitudinal polarization [3]. Accordingly, following the release of the initial polarization, the coherent LO phonon emits a terahertz electromagnetic wave. Thus, the dynamically oscillating polarization due to the coherent LO phonon is directly observed in terahertz time-domain spectroscopy. This is advantageous because, in the pump-probe measurement, the motion of the coherent phonon is indirectly observed through the modulation of the dielectric function. In earlier work, we observed that (11*n*)-oriented GaAs/$In_{0.1}Al_{0.9}As$ strained multiple quantum wells (MQWs) emit an intense quasi-monochromatic terahertz wave from the coherent GaAs-like LO phonon through the release of piezoelectric fields in the GaAs well layers by photogenerated carriers [5]. Here, we settle an issue whether the terahertz-wave emission from the coherent GaAs-like LO phonon is affected by screening effects of high density photogenerated carriers.

In the present work, we investigated the screening effect of photogenerated carriers on the terahertz waves from the coherent GaAs-like LO phonons in the (11*n*)-oriented GaAs/$In_{0.1}Al_{0.9}As$ strained MQWs. We found that the amplitude of the terahertz wave from the coherent GaAs-like LO phonon increases as the orientation index approaches $n = 1$, which enhances the piezoelectric field. The Fourier transform spectra of the terahertz waveforms show an intense quasi-monochromatic band of the coherent GaAs-like LO phonon. We found that the integrated intensity of the coherent GaAs-like LO-phonon band tends to saturate with an increase in the pump power. We analyzed the saturation behavior in terms of the unit-power intensity corresponding to excitation efficiency of the terahertz wave.

2. Samples and Experimental Procedure

The present samples, the GaAs/$In_{0.1}Al_{0.9}As$ strained MQWs, were grown on (11*n*)-oriented semi-insulating GaAs substrates by molecular beam epitaxy. The orientation indices *n* were 2, 3, and 4. The MQW structure of each sample consisted of 20 periods of the 10-nm-thick GaAs well layer and 10-nm-thick $In_{0.1}Al_{0.9}As$ barrier layer. We simply refer the (11*n*)-oriented GaAs/$In_{0.1}Al_{0.9}As$ strained MQW as the (11*n*) MQW. We note that an $In_{0.1}Al_{0.9}As$ metamorphic buffer layer with a thickness of 1.0 μm was grown in each sample. Accordingly, only the GaAs layer is influenced by the biaxial stain resulting from the lattice mismatch between GaAs and $In_{0.1}Al_{0.9}As$. This biaxial strain produces the piezoelectric field in the GaAs layer. The lattice constants of GaAs and $In_{0.1}Al_{0.9}As$ are 0.5633 and 0.5701 nm, respectively [6]. Using these values, the biaxial tensile strain ε in the GaAs layer is calculated to be 0.84%. Consequently, the piezoelectric fields in the GaAs layers of the (112), (113), and (114) MQWs were calculated to be 144, 72, and 41 kV/cm, respectively, according to the theory in Ref. 7. We notice that the value of 144 kV/cm is about one-third of the breakdown electric field in GaAs [8]. The calculated piezoelectric field strength increases as the orientation index approaches $n = 1$. This is reasonable because the GaAs crystal has a polar axis along the [111] direction. The terahertz time-domain measurement was performed at room temperature, using a Ti:sapphire femtosecond pulse laser. The duration time and repetition of the laser pulse were about 50 fs and 90 MHz, respectively. The pump beam was focused on the sample with an incidence angle of 45°. The diameter of the beam spot on the sample surface was about 100 μm. The emitted terahertz wave was collected with the use of two off-axis parabolic mirrors, and was detected by an optically gated photoconductive dipole antenna with a gap of 6.0 μm formed on a low-temperature-grown GaAs layer. The peak photon energies of the pump and probe beams were the same value: 1.55 eV. We used a purge of dry nitrogen gas and kept the humidity being ~5% to suppress the noise arising from the absorption of water vapor. The details of the experimental setup are described in Ref. 9.

3. Experimental Results and Discussion

Figure 1 shows the terahertz waveform of each sample at a pump power of 70 mW. The bipolar pulse, which originates from the ultrafast current of the photogenerated carriers [10], appears around the time delay of 0 ps. The bipolar pulse accompanies with the long-lived oscillating pattern with a period of 115 fs. Note that the oscillation amplitude increases with the change in the orientation index toward to $n = 1$. In order to analyze the terahertz waveform, we applied the Fourier transform to the terahertz waveform. The Fourier power spectrum of each sample is shown in Fig. 2. The negligibly weak signal in the frequency range from 0 to 5 THz originates from the bipolar pulse around the time delay of 0 ps. In the frequency range from 8 to 10 THz, the intense sharp band, which has a peak at the frequency of 8.7 THz, appears in each sample. This band is attributed to the coherent GaAs-like LO phonon. The peak frequency of the GaAs-like LO phonon band is about 0.1 THz lower than the LO-phonon frequency (8.8 THz) in a GaAs single crystal. The value of the 0.1 THz corresponds to the biaxial strain of about -0.9% because the strain-induced frequency shift $\Delta\omega$ of the GaAs-like LO phonon is in the following relation with ε: $\Delta\omega \cong -11\varepsilon$ (THz) [11]. This phenomenon indicates the presence of the biaxial tensile strain in the GaAs layer. The intensity of the coherent GaAs-like LO-phonon band is markedly increased as the orientation index approaches $n = 1$. Since the approach of *n*

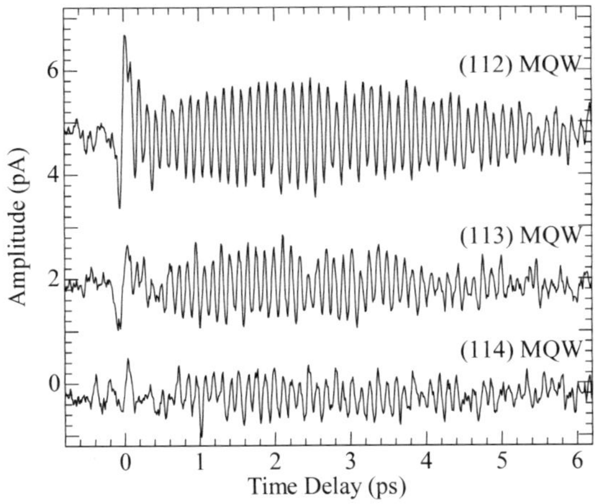

Figure 1: Amplitudes of the terahertz waveforms of the (114), (113), and (112) MQWs as a function of time delay at room temperature. For clarity, each waveform is vertically shifted.

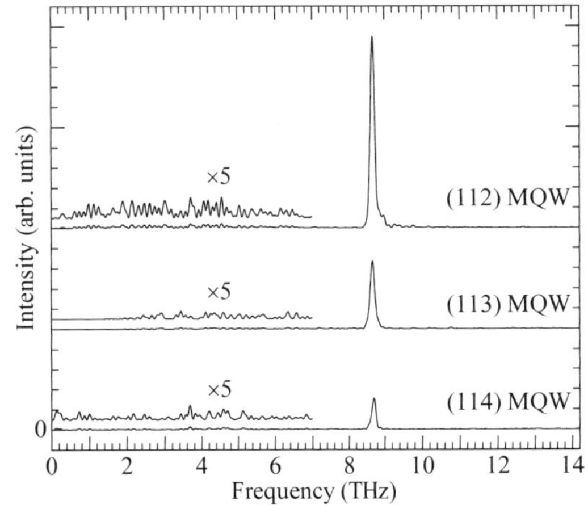

Figure 2: Fourier power spectra of the terahertz waveforms shown in Fig.1.

to 1 causes the enhancement of the piezoelectric field, the increase in the intensity of the coherent GaAs-like LO-phonon band originates from the piezoelectric field enhancement, which enlarges the initial polarization of the coherent GaAs-like LO phonon.

In order to investigate how the screening effect influences the coherent GaAs-like LO phonon, we measured the terahertz waves from the (11*n*) MQWs at various pump powers. Figure 3 shows the integrated intensity of the coherent GaAs-like LO-phonon band as a function of pump power P_{pump}. The integrated intensity monotonically increases with an increase in the pump power; however the pump-power dependence exhibits a saturation behavior. We evaluated the saturation behavior as follows. The intensity of the coherent GaAs-like LO-phonon band I_{LO} is proportional to the square of pump power in principal [12]. Thus, we introduce the parameter of the unit-power density defined as $I_{\text{LO}}^{\text{unit}} = I_{\text{LO}}/P_{\text{pump}}^{2}$. The unit-power intensity corresponds to the excitation efficiency of the terahertz wave from the coherent LO phonon. In Fig. 4, we plotted the unit-power intensity of each sample as a function of pump power. The unit-power intensity monotonically decreases with an increase in the pump power in each sample. The increase in the pump power leads to an increase in photogenerated

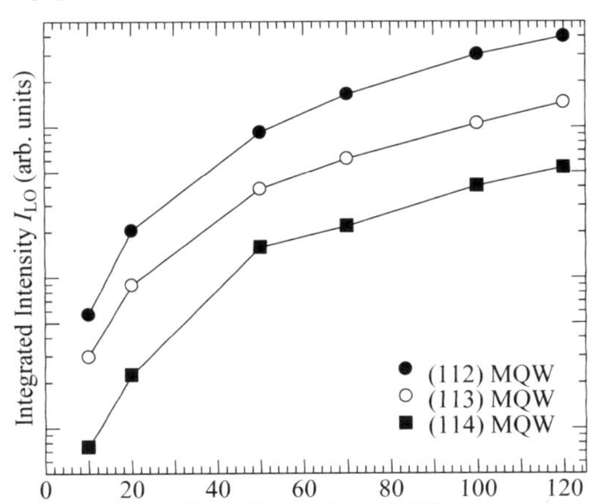

Figure 3: Integrated intensity of the coherent GaAs-like LO-phonon band as a function of pump power. The solid lines are the guides for eyes.

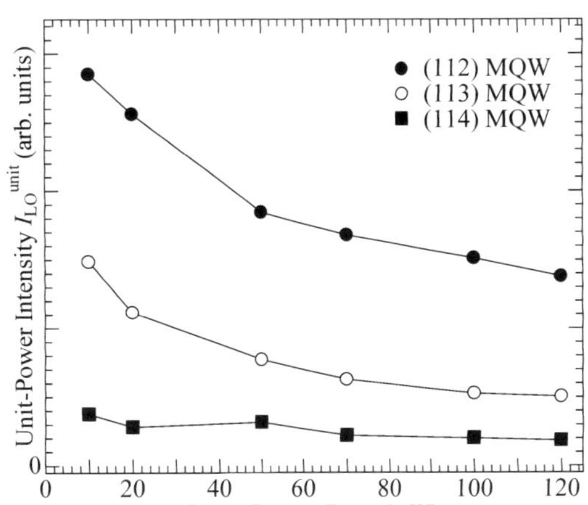

Figure 4: Unit-power intensity of the coherent GaAs-like LO-phonon band as a function of pump power. The solid lines are the guides for eyes.

carriers the GaAs layer, which results in enhancement of the Coulomb screening effect on the piezoelectric field. In other words, the Coulomb screening effect lowers the piezoelectric field, which reduces the initial polarization of the coherent GaAs-like LO phonon. Consequently, the Coulomb screening is attributed to the origin of the saturation behavior of the pump-power dependence of the coherent GaAs-like LO-phonon intensity in the (11*n*) MQWs.

4. Summary

We have investigated the screening effect of photogenerated carriers on the terahertz radiation from coherent GaAs-like LO phonons in the (11*n*)-oriented GaAs/$In_{0.1}Al_{0.9}As$ strained MQWs. We have observed the intense quasi-monochromatic terahertz wave from the coherent GaAs-like LO phonon, the initial polarization of which is enhanced by the strong piezoelectric field. The intensity of the coherent GaAs-like LO-phonon band exhibits a saturation behavior with an increase in the pump power. We have evaluated the excitation efficiency of the terahertz wave from the coherent GaAs-like LO phonon with the use of the unit-power intensity. From the pump-power dependence of the unit-power intensity, we conclude that the considerable screening effects of high-density photogenerated carriers on the piezoelectric field cause the saturation of the intensity of the terahertz wave from the coherent GaAs-like LO phonon.

References

[1] Dekorsy T, Cho G C and Kurtz H 2005 Coherent Phonon in Condensed Media *Light Scattering in Solids VIII*, ed M Cardona and G Güntherodt (Berlin: Springer) Chapter 4.
[2] Cho G C, Kütt W and Kurz H 1990 *Phys. Rev. Lett.* **65** 764.
[3] Pfiffer T, Decorsy T, Kütt W and Kurtz W 1992 *Appl. Phys.* A **55** 482.
[4] Gu P and Tani M 2005 Terahertz Radiation from Semiconductor Surfaces *Terahertz Optoelectronics*, K Sakai ed. (Berlin: Springer) pp 63-97.
[5] Takeuchi H, Asai S, Tsuruta S and Nakayama M 2012 *Appl. Phys. Lett.* **100** 242107.
[6] Madelung O 1996 *Semiconductors -Basic Data* (Berlin: Springer). The lattice constant of $In_{0.1}Al_{0.9}As$ is estimated with use of a linear interpolation between those of InAs and AlAs.
[7] Anastassakis E 1992 *Phys. Rev.* B **46** 4744.
[8] Grove A S 1967 *Physics and Tecnology of Semiconductor Devices* (New York: Wiley) p 103.
[9] Takeuchi H, Sumioka T and Nakayama M 2017 *J. Vac. Sci. Technol.* A **35** 04D104.
[10] Zhang X –C and Auston D 1992 *J. Appl. Phys.* **71** 326.
[11] Nakayama M, Kubota K, Kanata T, Kato H, Chika S and Sano N 1985 *J. Appl. Phys.* **58** 4342.
[12] Mizoguchi K, Mizumoto A, Nakayama M, Saito S, Syouji A, Sakai K, Yamamoto N and Akahane K 2006 *J. Appl. Phys.* **100** 103527.

IOP Conf. Series: Journal of Physics: Conf. Series **906** (2017) 012021 doi:10.1088/1742-6596/906/1/012021

Thermoelectric current in tubular nanowires in transverse electric and magnetic fields

Gunnar Thorgilsson, Sigurdur I. Erlingsson, and Andrei Manolescu
School of Science and Engineering, Reykjavik University, Menntavegi 1, IS-101 Reykjavik, Iceland
E-mail: gunnarth@ru.is

Abstract. In the presence of a transverse magnetic field, the charge current in nanowires can flow from the hot to the cold reservoir, but also backwards. The sign change can be obtained by increasing the temperature bias or the magnetic field. This behavior occurs when the magnetic field is sufficiently strong. Here, we will investigate how the size of the anomalous backward-flowing current is affected by an electric field perpendicular to the nanowire. The interplay of the electric and magnetic field modifies the dispersion curves, which will show up in the transport properties. We will also investigate how the presence of impurities affects the anomalous current. The electric field affects backscattering due to impurities, and thus the thermoelectric current reversal. Preliminary results show that the current reversal can survive in the presence of impurities.

1. Introduction

A temperature gradient across a conducing material induces an energy gradient, which in turn results in particle transport. In an open circuit, where no net current flows, a voltage is then generated when two ends of a sample are maintained at different temperatures — this is the Seebeck effect and the linear voltage response is known as the thermopower. The hotter particles have larger average kinetic energy, and the net particle flow is therefore generally from the hot to the cold side. The thermopower and thermoelectric current can be positive or negative, depending on the type of charge carriers, i.e., electrons or holes.

In comparison to this macroscopic case, the thermopower at the nanoscale has special characteristics. For example, if the energy separation between the quantum states of the system is larger than the thermal energy, the thermopower may alternate between positive and negative values, depending on the position of the Fermi level relative to a resonant energy, which can be controlled with a gate voltage. These oscillations were predicted a long time ago [1], and subsequently experimentally observed in quantum dots [2, 3]. In these examples, the charge carriers are electrons and the sign change of the thermopower means that they travel from the cold side to the hot side, which may appear counterintuitive.

Observing negative thermopower at the nanoscale is difficult for two reasons: the currents are small ($\sim$ pA [4]) and it is hard to maintain a large temperature difference across short distances [5]. However, cylindrical nanowires are well suited for observing negative thermopower since their length can reach μm and the currents are of the order of tens of nA [6]. Here, we focus on how an electric field affects the thermoelectric current.

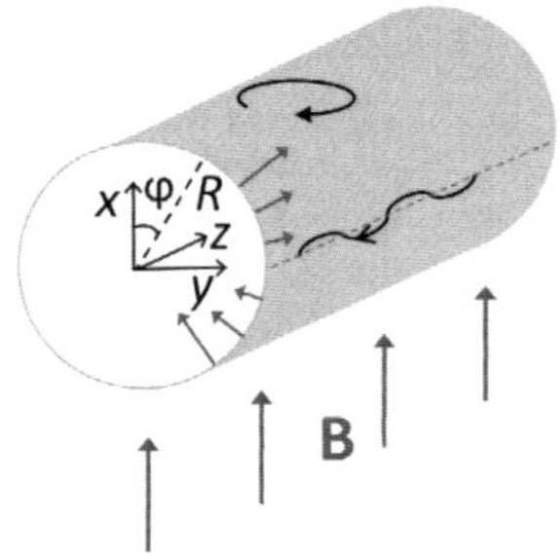

Figure 1. A hollow cylindrical nanowire (light green) in a uniform transverse magnetic field (thick red arrows). On the top and bottom regions of the cylinder, electrons move in closed cyclotronic loops, whereas on the lateral sides longitudinal snaking orbits are formed along the lines where the radial projection of the magnetic field (thin red arrows) is zero.

2. Model and methodology

We model a cylindrical nanowire [7, 8] as a two-dimensional sheet wrapped into a cylinder of radius R. We choose the coordinate system such that magnetic field is along the x-axis, $\mathbf{B} = (B, 0, 0)$, the vector potential being $\mathbf{A} = (0, 0, By) = (0, 0, BR\sin\varphi)$. The electric field is also applied along the x axis, such that $\mathcal{E} = (\mathcal{E}_x, 0, 0)$. In this setup, the states are affected by a competition of the electric and magnetic fields. An electron traveling in the positive z-direction will tend to be pulled towards $\varphi = -\pi/2$ by the Lorentz force and an electron travelling in the opposite direction will tend to be pulled to the opposite side of the cylinder, i.e., $\varphi = \pi/2$ (see Fig. 1). The resulting states are known as snaking states [9]. The electric field tends to pull electrons toward $\varphi = 0$, which will affect the number of electrons in the counterpropagating snaking states. As we will see later, this will affect the transport properties in the presence of impurities. The clean nanowire, i.e., in the absence of impurities, is translationally invariant along the z direction and the momomentum operator p_z can be replaced by is eigenvalue $\hbar k$. The Hamiltonian can be written as

$$H(k) = \frac{\hbar^2}{2m_{\text{eff}}R^2}\left[-\frac{\partial^2}{\partial\varphi^2} + \left(Rk + \frac{eBR^2}{\hbar}\sin\varphi\right)^2\right] + e\mathcal{E}_x R\cos\varphi - \frac{g_{\text{eff}}\mu_{\text{B}}}{2}B\sigma. \quad (1)$$

In this example, we consider material parameters for GaAs, i.e., the effective mass $m_{\text{eff}} = 0.066\,m_0$ and a g-factor $g_{\text{eff}} = -0.44$, μ_{B} being the Bohr magneton and $\sigma = \pm\mathbb{Z}$ being the spin label. For $B = 0$ and $\mathcal{E}_x = 0$, the angular part of the Hamiltonian has eigenfunctions $e^{i\varphi n}/\sqrt{2\pi}$, $n \in Z$, and the single-electron energy dispersions are ordinary parabolas as a function of k. These eigenfunctions define a basis set, $|nk\sigma\rangle$, which we use for $B \neq 0$ to diagonalize numerically (1). The convergence is reached with $|n| \leq 50$. It is convenient to introduce a length scale $\ell_E = \left(\hbar^2/\left(2me\mathcal{E}_x\right)\right)^{1/3}$ associated with the electric field strength $\mathcal{E}_x$. When $l_E \gg R$, the effect of the electric field is small, but once $\ell_E \lesssim R$, the electric field starts playing a role. Similar arguments hold for the the magnetic length $\ell_c = \sqrt{eB/\hbar}$, i.e., when $\ell_c \lesssim R$, the magnetic field becomes important.

The charge current through the nanowire, driven by a temperature gradient, can be calculated using the Landauer formula

$$I_c = \frac{e}{h}\int T(E)\left[f_{\text{R}}(E) - f_{\text{L}}(E)\right]dE, \quad (2)$$

where $f_{\text{L/R}}(E)$ are the Fermi functions for the left/right reservoir with chemical potentials $\mu_{\text{L/R}}$ and temperatures $T_{\text{L/R}}$. We consider a temperature bias, $T_{\text{R}} > T_{\text{L}}$, beyond linear response, and no potential bias, $\mu_{\text{L}} = \mu_{\text{R}} = \mu$.

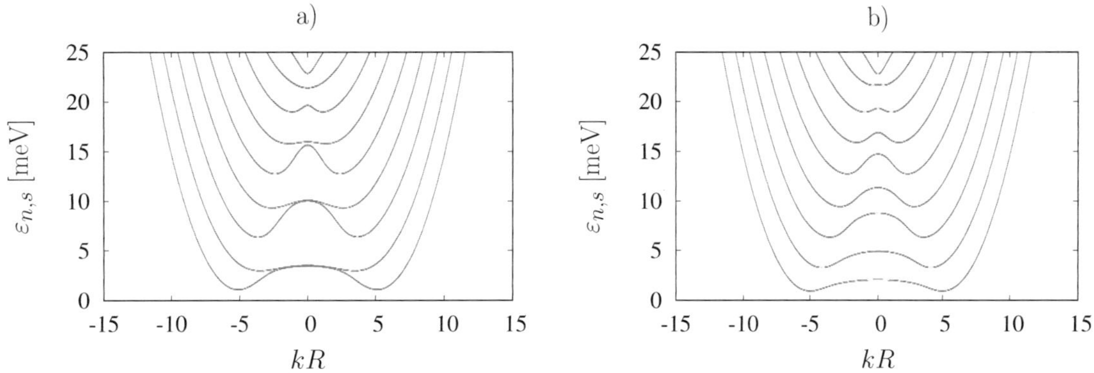

Figure 2. The energy spectra for a cylindrical nanowire with $R = 30\,\text{nm}$ and $B = 4.0\,\text{T}$. The spectrum for a weak electric field $\ell_E = 3.0R$ a) and strong magnetic field $\ell_E = 0.75R$ b). Degenerate cyclotron states at $k = 0$ in a) are split by the electric field as seen in b).

Realistic nanowires are never totally ballistic, but state-of-the-art nanowires are quasi-ballistic in the sense that transport measurements do show conductance steps [8, 10]. In order to model transport in the presence of impurities, we will introduce an impurity potential $V_{\text{imp}}(z, \varphi) = \sum_i W\delta(z - z_i)\delta(\varphi - \varphi_i)$, where W is the impurity strength. We consider a fixed impurity configuration, i.e., no ensemble average. To some extent the results depend on the impurity configuration, as also seen in experiments. There, the conductance can show complicated, but reproducible behavior for a given nanowire, whereas the conductance for *another* nanowire will yield conductance whose structure (position of peaks, etc.) will be different [11], but reproducible as well. The transmission function in the case when impurities are included is obtained using the recursive Green's function method [12, 6].

3. Results and discussion

In Fig. 2, a typical spectrum for the system is shown for $R = 30\,\text{nm}$ and magnetic field $B = 4.0\,\text{T}$. A weak electric field $\ell_E = 3.0R$ in a) leaves the cyclotron and snaking states relatively unaffected but a stronger electric field, corresponding to $\ell_E = 0.75R$ in b), splits the states at $kR = 0$. Since the electric field is parallel to the magnetic field, the cyclotron states residing on the top and bottom of the nanowire get split by $\mathcal{E}_x$, while the snaking states are mostly unchanged as the electrostatic potential is zero around $\varphi = \pm\pi/2$. The local magnetic field on the cylinder surface is determined by the orientation of the normal (which always points outwards) relative to the magnetic field (Fig. 1).

For given properties of the reservoirs ($\mu_{L,R}$ and $T_{L,R}$), the current is fully determined by the transmission function [12]. In the case of a ballistic wire, the transmission function can be found by simply counting the number of propagating modes. The important point here is that, due to the coexistance of cyclotron and snaking states the dispersion curve minima occur at non zero kR. Looking at the first energy band, one sees that, between the cycloton maxima and snaking-state minima, there will be two propagating modes; when accounting for spin the transmission should be via $2 \times 2 = 4$ modes, which drops to $2 \times 1 = 2$ for energies above the cyclotron maximum at $kR = 0$. It is this drop in conductance, as a function of energy, that gives rise to the anomalous current [6]. The difference is Fermi functions is an odd function of energy around μ, so placing μ where $T(E)$ drop, will lead to a negative (*anomalous*) current.

Figures 3 a) and 3 b) show the transmission function T and the charge current I_c versus temperature of the right contact, respectively, for $l_E = 3.0R$ (weak electric field) and for varying density of impurities, n_i. Analogous plots for a stronger electric field, $l_E = 0.75R$, are shown in Figures 3 c) and 3 d). The effect of the electric field is to split the cyclotron states at $k = 0$. The

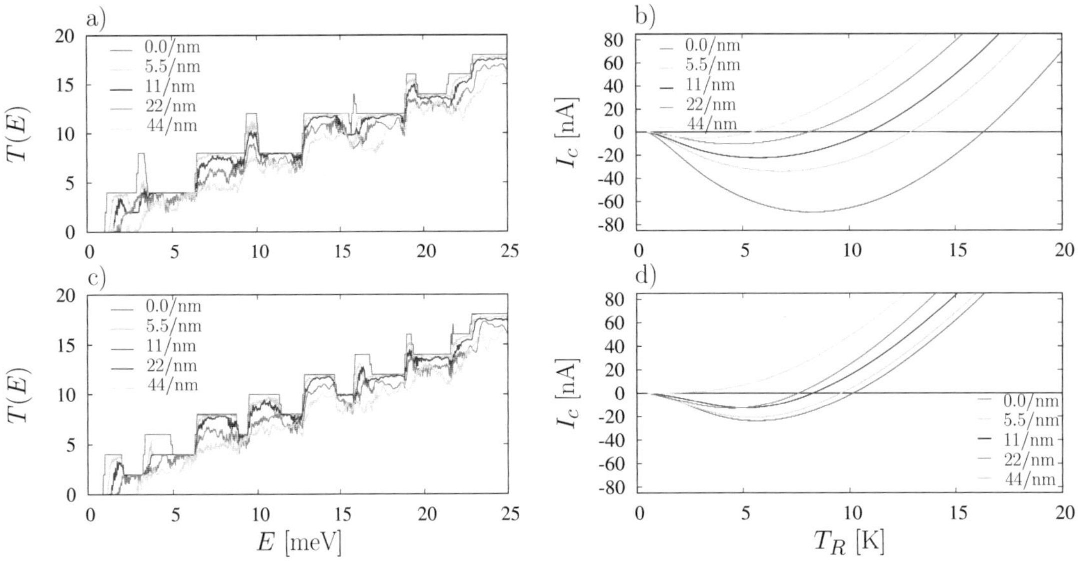

Figure 3. For the weak electric field case $\ell_E = 3.0$, a) the transmission function T and b) charge current I_c are calculated for different impurity densities n_i. The anomalous current is clear and is in the range of 60 nA. In c) and d) the corresponding results for the strong electric field, $\ell_E = 0.75$, are shown. The lowest snaking states tend to get washed out more readily in the presence of a strong electric field.

cyclotron states that get lowered in energy are brought closer to the snaking states, making it easier to scatter from $k \to -k$. The impuries quickly reduce the *anomalous* structure in $T(E)$, eventually leading to full suppression at $n_i = 44$, see Fig. 3 d).

In conclusion, the applied electric field tends to suppress the snaking states, which in turn leads to a reduced *anomalous* current. The electric field also leads to increased effectiveness of electron backscattering from impurities. The electric field can thus be used to test the properties of the snaking states and the resulting current reversal.

This work is supported by the research funds of Reykjavik University.

References

[1] Beenakker C W J and Staring A A M 1992 *Phys. Rev. B* **46**(15) 9667–9676
[2] Staring A A M, Molenkamp L W, Alphenaar B W, van Houten H, Buyk O J A, Mabesoone M A A, Beenakker C W J and Foxon C T 1993 *Europhys. Lett.* **22** 57
[3] Dzurak A, Smith C, Pepper M, Ritchie D, Frost J, Jones G and Hasko D 1993 *Solid State Commun.* **87** 1145 – 1149 ISSN 0038-1098
[4] Svensson S F, Hoffmann E A, Nakpathomkun N, Wu P, Xu H Q, Nilsson H A, Sanchez D, Kashcheyevs V, and Linke H 2013 *New J. Phys.* **15** 105011 .
[5] Roddaro S, Ercolani D, Safeen M A et al. 2014 *Nano Res.* **7** 579
[6] Erlingsson S, Manolescu A, Nemnes G, Bardarson J and Sanchez D 2017 *Phys. Rev. Lett.* **119**(3) 036804
[7] Gül O, Demarina N, Blömers C, Rieger T, Lüth H, Lepsa M I, Grützmacher D and Schäpers T 2014 *Phys. Rev. B* **89**(4) 045417
[8] Heedt S, Manolescu A, Nemnes G A, Prost W, Schubert J, Grützmacher D and Schäpers T 2016 *Nano Lett.* **16** 4569–4575
[9] Manolescu A, Rosdahl T, Erlingsson S, Serra L and Gudmundsson V 2013 *Eur. Phys. J. B* **86** 445 ISSN 1434-6028
[10] Heedt S, Traverso Ziani N, Crépin F, Prost W, Trellenkamp S, Schubert J, Grützmacher D, Trauzettel B and Schäpers T 2017 *arXiv:1701.08439*
[11] Wu P M, Gooth J, Zianni X, Svensson S F, Gluschke J G, Dick K A, Thelander C, Nielsch K and Linke H 2013 *Nano Lett.* **13** 4080–4086
[12] Ferry D K and Goodnick S M 1997 *Transport in Nanostructures* (Cambridge: Cambridge University Press)

IOP Conf. Series: Journal of Physics: Conf. Series **906** (2017) 012022 doi:10.1088/1742-6596/906/1/012022

Optoelectronic and photoacoustic studies of an organic dye synthesized through green route

S Vijayakumar[1,3,*], S Sreelatha[2], M Hatamimoslehabadi[3], C S Yelleswarappu[3]

[1] Department of Physics, N. S. S. College Pandalam, Kerala-689501, India
[2] Department of Chemistry, N. S. S. College Pandalam, Kerala-689501, India
[3] Department of Physics, University of Massachusetts Boston, MA 02125, USA

*v.sadasivannair@umb.edu, jevijay@gmail.com

Abstract. An azo dye was prepared through an environmentally benign and economically feasible synthesis route with cardanol as a starting material. Cardanol is a cost-effective and renewable natural source obtained from Cashew Nut Shell Liquid, a by-product of the cashew industry. The dye was spectrally characterized by IR, UV-Vis, NMR and fluorescence studies. UV-Vis absorption showed a bathochromic shift between solvents of lower and higher polarities. Nonlinear optical and photoacoustic properties were studied using a frequency doubled Nd:YAG laser producing 532 nm laser pulses of 3 ns pulse width. Results show that the nonlinear absorption coefficient decreases with the increase of on-axis intensity, suggesting excited state absorption as the principal mechanism. The observed nonlinearity has applications in optoelectronics.

1. Introduction

The rapid development of photonic technology demands new materials of precisely controlled chemical and structural properties [1]. Organic materials, specifically azobenzene molecules have gained wide interest due to their promising uses in many fields such as optoelectronics, biomedical studies, liquid crystalline displays, and pharmaceutical industry [2]. They absorbs light in the visible spectrum (400-700 nm) and exhibit resonance of electrons, which is a stabilizing force in organic compounds. The nonlinearity of these molecules has been found to originate from a strong delocalization of π-electrons along the length of the molecule [3]. Apart from the established applications of azobenzene molecules recently they have recently been considered as an alternative material for use in dye-sensitized solar cells [4]. Also, azobenzene derivatives are effectively used as smart targeted carrier for drug delivery systems [5]. However, when it comes to the choice of sensitizer synthesis, environmental and toxicological concerns, as well as economic feasibility should be taken into consideration. Motivated by these concerns, we synthesized the azobenzene compound cardanol as a starting material. Cardanol, a mixture of meta alkylphenols is obtained from Cashew Nut Shell Liquid (CNSL), a by-product of the cashew industry. Optimization of photophysical and chemical characteristics along with independent measurement of both nonlinear optical and photoacoustic properties is important to identify such materials for practical applications. In this paper we report the spectral, nonlinear optical and photoacoustic properties of an azo dye, 4-(2-carboxyl-phenylazo)-3-[pentadec-8-enyl] phenol synthesized using green-chemistry techniques. The results show that the dye is suitable for optoelectronic applications and its photoacoustic response can be further exploited as a contrast agent in photoacoustic imaging.

IOP Conf. Series: Journal of Physics: Conf. Series **906** (2017) 012022 doi:10.1088/1742-6596/906/1/012022

2. Methods

The prepared dye was purified by column chromatography [hexane:chloroform(1:1)] and was spectrally characterized by IR, UV-Vis, NMR and fluorescence studies. For nonlinear optical and photoacoustic studies, we used novel photoacoustic z-scan (PAZ-scan) technique [6] which enables concurrent monitoring of both the optical transmission and photoacoustic response of the sample under investigation. A frequency doubled Nd:YAG laser (Minilite II, Continuum) is focused onto the sample with a 20 cm focal length lens. The sample is placed in a 2 mm quartz cuvette and mounted in a custom made cell that contains water for ultrasound coupling. The sample cell is placed at 45° with respect to the incident laser beam and the effective optical path length equal to 2.83 mm. The whole system is mounted on a XYZ translation stage (Thorlabs NRT 150) and is moved such that the focal point of the beam scans the sample along its length (in the z direction). As the laser pulse is incident on the sample, some of the energy delivered is absorbed and converted into heat. This produces pressure transients and wideband ultrasonic emission which is detected using a 10 MHz focused water immersion transducer (Olympus NDT Inc.).

3. Results

The dye was synthesized in two steps namely diazotization using anthranilic acid followed by coupling with cardanol (figure 1). The pH of the diazonium solution was adjusted to 6.5 by the addition of an appropriate amount of 1M sodium bicarbonate solution. The precooled solution of cardanol in ethanol was injected drop wise to the above pH monitored solution at 0°C. The reaction mixture was stirred at 0°C for 40 minutes by which time the product was formed as an orange viscous liquid.

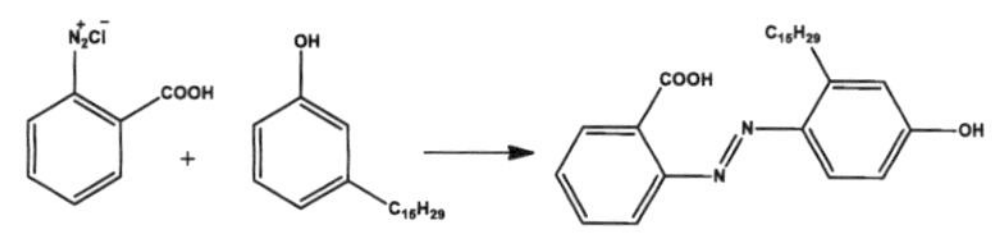

Figure 1. Coupling of anthanilic acid hydrochloride with cardanol.

The optical absorption and fluorescence emission spectra of the dye is shown in the figure 2. The UV-Vis studies of the dye in different solvents hexane and chloroform indicates that the dye exhibited $\pi \rightarrow \pi^*$ transition and showed a bathochromic shift when switching over from less polar to more polar solvents. This depends on the nature of the electron donating and accepting groups. In the case of $\pi \rightarrow \pi^*$ transition, the excited state is more polar than the ground state and the dipole-dipole interaction with solvent molecules lower the energy of the excited state more than that of the ground state. Therefore a polar solvent decreases the energy of $\pi \rightarrow \pi^*$ transition and absorption maxima appear red shifted from hexane to chloroform solvent. The fluorescence emission spectra of the dye was recorded in chloroform at 293 K excited at 350-400 nm and a high intensity was recorded in the range ~600-700 nm. In accordance with these results the material is suitable for using as a fluorescent marker and light emitter in organic light emitting diodes.

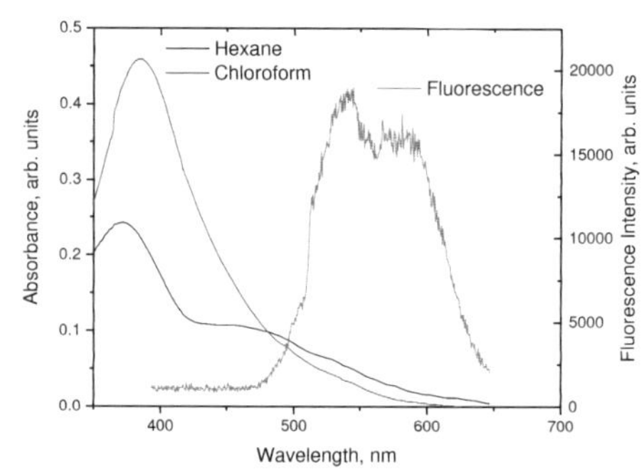

Figure 2. UV –Vis and Fluorescent emission spectra of the dye.

For NLO and PA studies the dye is dissolved in chloroform with a concentration of 1×10^{-4}mol/mL. The normalized open aperture (OA), closed aperture (CA) Z-scan curves and PA curve at on-axis peak intensity 1.16×10^{13} W/m^2 is shown in figure 3. As the intensity of the incident light changes, the optical transmittance varied in a way which depends on the nonlinear absorption properties of the sample. When an aperture is inserted prior to the output detector, the phase variation will induce a change in energy transmittance through the aperture. The CA Z-scan curve shows this variation and the dye exhibits a self-defocussing effect which leads to negative nonlinear refractive index.

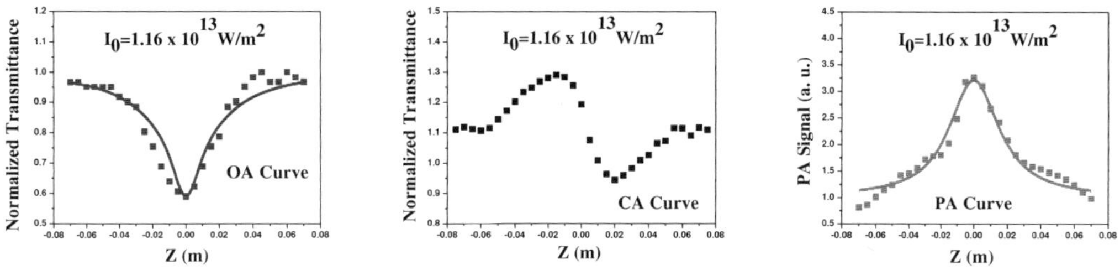

Figure 3. Optical and PAZ-scan curves. The solid lines are theoretical fits of experimental data.

For a temporally Gaussian pulse, when all the transmitted energy is collected, the OA normalized transmittance is given by,

$$T(Z) = \sum_{m=0}^{m=\infty} \frac{\left(-\beta I(Z)L_{eff}\right)^m}{(m+1)^{3/2}}, \; L_{eff} = \frac{1-e^{-\alpha L}}{\alpha} \quad (1)$$

where β is the nonlinear absorption coefficient, I(Z) is the incident laser intensity, L is the sample length and α is the linear absorption coefficient. Fitting the OA data with equation (1) gives a β value which in the present study is found to be 5.80×10^{-10} m/W. Again, when light energy is absorbed by the sample and converted in to heat, thermal expansion of the media arises and acoustic waves (PA signal) are formed. The generated acoustic waves can be given as [6],

$$p = \Gamma \mu_a I(Z) \quad (2)$$

where μ_a is the optical absorption coefficient of the sample and Γ is the Gruneisen parameter. Fitting the PA experimental data with equation (2), also measured the nonlinear absorption properties of the sample. The obtained value of β in this case is 5.23×10^{-10} m/W in agreement with the NLO measurement. Comparison of the reported studies of the nonlinear optical properties of novel azo benzenes [7], suggests that the dyes possess large third order nonlinear optical properties. In addition, the dye shows good photoacoustic response which suggests their applicability as contrast agents in PA imaging.

To understand the mechanism behind the observed nonlinear absorption, the excited state absorption cross-section (σ_{ex}) of the sample was measured from the open-aperture Z-scan data [8]. For this OA data is fitted to the transmittance equation,

$$T = ln\left(1+\frac{q_0}{1+x^2}\right) \Big/ \left(\frac{q_0}{1+x^2}\right), \; q_0 = \frac{\sigma_{ex}\alpha F_0 L_{eff}}{2\hbar\omega} \quad (3)$$

F_0 is the fluence of the laser at the focus, h is the Planck's constant, ω is the angular frequency of the laser, $x=Z/Z_R$ where Z is distance of the sample from the focus with a positive value between focus and detector and Z_R is the Rayleigh length. The ground state absorption cross-section, σ_g is calculated from the equation, $\alpha=\sigma_g N_a C$; where N_a is the Avogadro's number and C is the concentration in moles/cm^3. The measured value of σ_{ex} is 1.55×10^{-18} cm^2 which is found to be larger than the value of σ_g, 1.32×10^{-19} cm^2. This is a favourable condition for reverse saturation absorption to occur. To confirm this, the value of β was evaluated with different on-axis peak intensities (I_0) ranging from 1.16×10^{13} W/m^2 to 1.70×10^{13} W/m^2 and found to decrease in value from 5.8×10^{-10} m/W to 3.21×10^{-10} m/W (figure 4). This result along with positive value of β

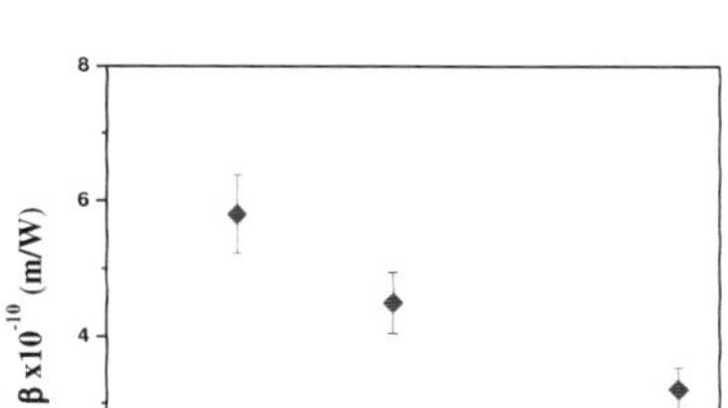

Figure 4. Variation of nonlinear absorption coefficient (β) with on-axis peak intensity (I_0).

IOP Conf. Series: Journal of Physics: Conf. Series **906** (2017) 012022 doi:10.1088/1742-6596/906/1/012022

suggests that excited state absorption be the principal mechanism for nonlinear absorption [9], which finds applications in optoelectronics.

4. Conclusion

An azo dye namely 4-(2-carboxyl-phenylazo)-3-[pentadec-8-enyl] phenol was prepared through green synthesis methods using cardanol as a starting material. Nonlinear optical, and photoacoustic studies were carried out at 532 nm with 3ns laser pulses using a photoacoustic Z-scan experiment. The material shows a self-defocussing nature and excited state absorption at 532 nm. The nonlinear absorption coefficient was estimated and found to be on the order of 10^{-10}m/W.

Acknowledgments
S Vijayakumar acknowledges the University Grants Commission, New Delhi, India for RAMAN Fellowship for Post Doctoral Research in USA.

References
[1] Moss D J, Morandotti R, Gaeta A L and Lipson M 2013 New CMOS-compatible platforms based on silicon nitride and Hydex for nonlinear optics *Nature Photonics* **7** 597-607
[2] Si G, Zhao Y, Leong E S P and Liu Y J 2014 Liquid-crystal-enabled active plasmonics: a review *Materials* **7** 1296-317
[3] Hales J M, Matichak J, Barlow S, Ohira S, Yesudas K, Brédas J-L, Perry J W and Marder S R 2010 Design of polymethine dyes with large third-order optical nonlinearities and loss figures of merit *Science* **327** 1485-8
[4] Zhang L and Cole J M 2014 TiO2-assisted photoisomerization of azo dyes using self-assembled monolayers: case study on para-methyl red towards solar-cell applications *ACS applied materials & interfaces* **6** 3742-9
[5] Wang Y, Li B, Zhang L, Song H and Zhang L 2012 Targeted delivery system based on magnetic mesoporous silica nanocomposites with light-controlled release character *ACS applied materials & interfaces* **5** 11-5
[6] Yelleswarapu C S and Kothapalli S-R 2010 Nonlinear photoacoustics for measuring the nonlinear optical absorption coefficient *Optics express* **18** 9020-5
[7] El Ouazzani H, Iliopoulos K, Pranaitis M, Krupka O, Smokal V, Kolendo A and Sahraoui B 2011 Second-and third-order nonlinearities of novel Push– Pull azobenzene polymers *The Journal of Physical Chemistry B* **115** 1944-9
[8] Wood G L, Miller M J and Mott A G 1995 Investigation of tetrabenzporphyrin by the Z-scan technique *Optics letters* **20** 973-5
[9] Couris S, Koudoumas E, Ruth A and Leach S 1995 Concentration and wavelength dependence of the effective third-order susceptibility and optical limiting of C60 in toluene solution *Journal of Physics B: Atomic, Molecular and Optical Physics* **28** 4537

Hydrodynamic modelling of terahertz rectification in AlGaN/GaN high electron mobility transistors

Juozas Vyšniauskas[1], **Alvydas Lisauskas**[1,2], **Maris Bauer**[2], **Dovilė Čibiraitė**[2], **Jonas Matukas**[1] and **Hartmut G Roskos**[2]

[1]*Radiophysics Department, Vilnius University, 10222 Vilnius, Lithuania*

[2]*Physikalisches Institut, Goethe-Universität Frankfurt, 60438 Frankfurt, Germany*

E-mail: alvydas.lisauskas@ff.vu.lt

Abstract. We report on the numerical modelling of rectification in a gated two-dimensional electron gas. We demonstrate that drift-diffusion-based and energy-relaxation-based models predict different features of rectified terahertz radiation as a function of gate bias. Whereas the widely accepted mechanism for rectification is considered to be plasmonic-based, there are conditions when diffusion currents originating by non-local carrier heating can dominate the response. Moreover, diffusive contributions can substantially enhance the response becoming an important phenomenon, which has to be considered in future designs of efficient transistor-based terahertz rectifiers.

The prediction of plasma wave-resonances [1] in sub-micrometer-long channel field-effect transistors, (FETs) triggered a strong interest in applied aspects of plasmonic effects, which promise the extension of otherwise standard device operation beyond the conventional cut-off frequency limits. The plasmons, relating to a wave-like propagation of charge carriers, can manifest already at microwaves, yet they become increasingly pronounced in the terahertz (THz) frequency range. By now, it has been numerously reported in experimental [2, 3] and theoretic [4, 5] studies that plasmonic rectification can be employed for efficient detection of THz radiation. Despite such progress, there are many unresolved questions left regarding the validity of predictions from existing hydrodynamic models when applied for the modeling of practical devices (rectifiers or detectors).

In this contribution, we present numerical modeling results of THz rectification in sub-micrometer size AlGaN/GaN high electron mobility transistors (HEMT). For our numerical calculations, we employ a commercial SYNOPSIS TCAD package allowing to solve coupled Poisson, drift-diffusion, continuity and energy relaxation equations in multi-dimensional (2D or 3D) geometries. These equations form the basis for the so-called quasi-hydrodynamic or energy-transport model, which differs from a full-hydrodynamic approach by the omission of momentum relaxation [6]. For the reference calculations (excluding carrier heating) we employ simple drift-diffusion equations, which are analogous to a distributed RC transmission line model [7].

Figure 1 presents the 2D profile of the structure. All numerical simulations are performed in 2D assuming homogeneous properties in lateral dimension set to be 1 μm wide. The 2D-electon gas is formed between the 25nm-thick $Al_{0.2}Ga_{0.8}N$ layer and the undoped (intrinsic) GaN substrate. The gate length is set to 300 nm whereas the ungated parts are selected to be slightly asymmetric: 300-nm-long at the source side and 500-nm-long at the drain side. The ungated regions are covered with a 25-nm-thick Si_3N_4 passivation layer. At the SiN/AlGaN interface we include $5\cdot 10^{13}$ cm^{-2} donors located 0.4 eV

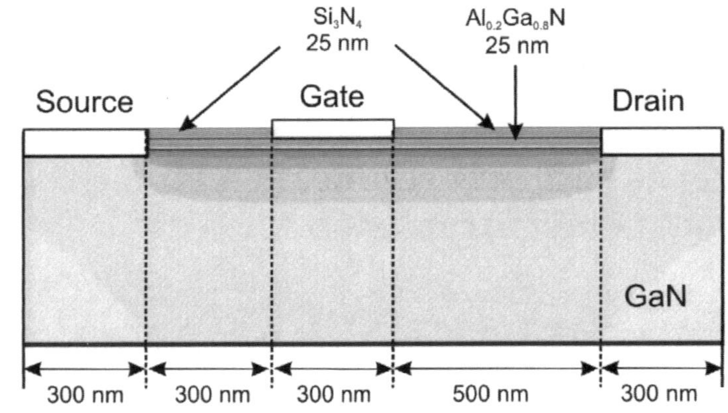

Figure 1. The structure of AlGaN/GaN HEMT employed for numerical simulations of THz rectification.

from the center of forbidden energy gap. Figure 2a shows the calculated static resistance as a function of gate bias. The knowledge of the static resistance allows estimation of the current responsivity in quasi-static conditions as it is usually performed for rectifying diodes [8]. $\mathfrak{R}_{I,QS} = 1/(2\sigma)d\sigma/dU_G = -1/(2R)dR/dU_G$. For the selected structure, the quasi-static current responsivity (presented in figure 2) peaks at -3.52 V reaching 5.47 A/W.

In order to calculate the rectification of THz radiation, we applied an oscillating voltage with an amplitude of 50 mV to the drain terminal placing the source terminal on ground. A fixed gate voltage implies that the gate terminal is equivalently shorted for all oscillating signals except of *DC*. Such a connection ensures asymmetric boundary conditions which are necessary for efficient rectification. We monitor the drain current in time domain which, after performing a Fast Fourier Transformation, allows obtaining the complex device impedance. The rectified signal is obtained from transient data by calculating the average current over one oscillation period after omitting the first 2 periods.

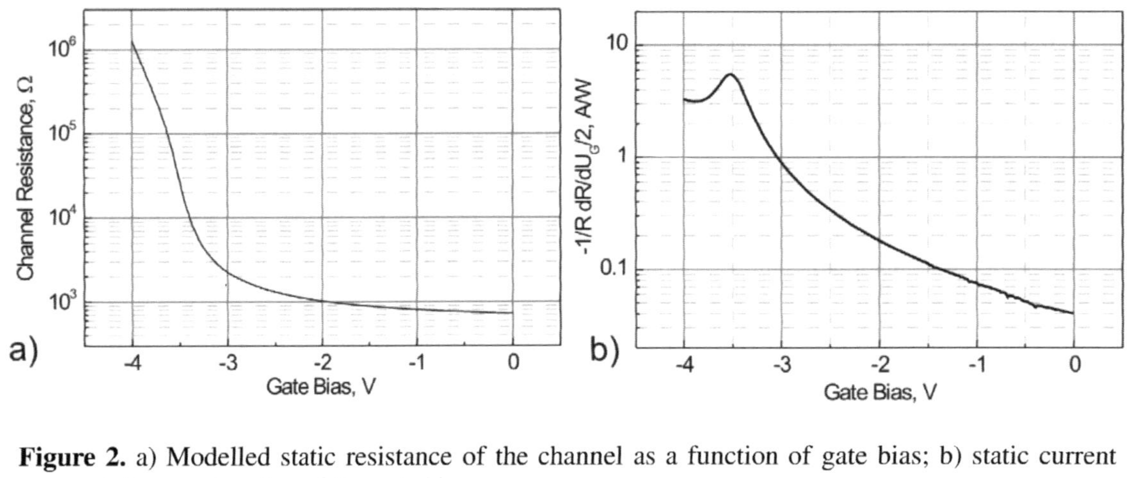

Figure 2. a) Modelled static resistance of the channel as a function of gate bias; b) static current responsivity as a function of the gate bias.

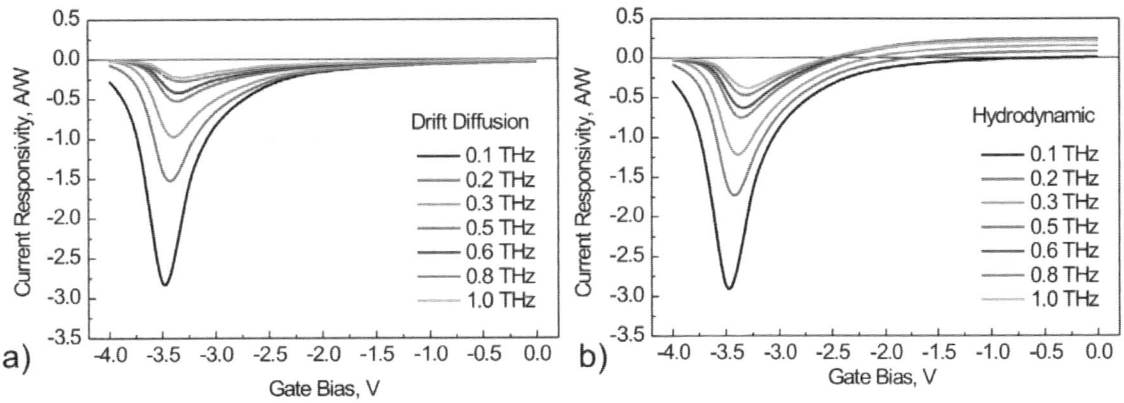

Figure 3. a) Current responsivity as a function of gate bias for various frequencies calculated with drift diffusion model, and b) calculated using hydrodynamic model.

We estimate the current responsivity by calculating the ratio between detected current and *AC* power absorbed by the device. Figure 3 presents the main simulation results. Since the oscillating signal is applied at the drain side, the current has negative sign. The left panel (a) presents the simulated responsivity using a drift-diffusion transport description whereas the right panel (b) presents modelling results when using the energy-transport hydrodynamic model. For comparatively low frequencies (100 GHz) both models predict similar rectification with a peak responsivity of about 2.8 A/W. With increasing frequency, the absolute value of the current responsivity decreases. Although intrinsic rectification in a distributed *RC* model should not be frequency dependent, such decrease originates from increasingly important power loss on the ungated parts at higher frequencies. It is important to note, that a hydrodynamic model predicts a reversal in the direction of the rectified current for large gate bias (in respect to threshold voltage) which cannot be explained by plasmon-based models. These predictions are supported with experimental data [9] of THz detection using AlGaN/GaN HEMTs with integrated broadband antennas.

Our experiments did not show this strong crossover between plasmonic mixing and diffusion of "warm" carriers, which might indicate that the omitted momentum relaxation terms lead to an overestimation of diffusive contributions. Moreover, the current responsivity at the maximum of the

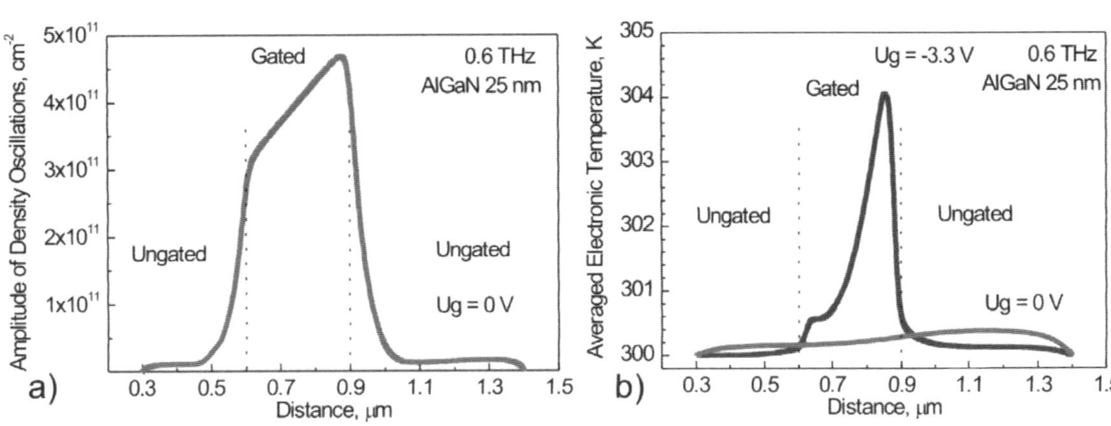

Figure 4. a) The lateral distribution of the amplitude of density oscillations using 600 GHz excitation for zero gate bias; b) Lateral distribution of averaged electronic temperature for two gate voltages.

absolute value of response is substantially increasing over drift-diffusion simulations at THz frequencies indicating the importance for thorough simulations of non-equilibrium carriers [10].

Figure 4 (a) demonstrates the distribution of the amplitude of the oscillating charge carriers for zero gate bias conditions, whereas the right panel (b) presents the distribution of average electronic temperature at zero gate bias and at gate bias resulting to the maximum of absolute current responsivity. Therefore, even if the carrier temperature does not change much, its lateral distribution results in an additional diffusion current, which at zero gate bias dominates the response with respect to plasmonic detection, whereas close to the threshold, the plasmonic detection is enhanced.

In summary, we show that heating of charge carriers by the coupled high-frequency radiation plays an important role in the modelled response of field-effect transistor-based THz detectors.

Acknowledgments
Frankfurt team acknowledges support from Hessian excellence program LOEWE, project "Terahertz-Kamera für die zivile Sicherheitstechnik". Vilnius team acknowledges support by the Research Council of Lithuania under the KITKAS project, contract LAT-04/2016.

References
[1] Dyakonov M and Shur M 1996 Detection, mixing, and frequency multiplication of terahertzradiation by two-dimensional electronic fluid *IEEE Trans. Electron. Dev.* **43** 380–387.
[2] Knap W *et al.* 2009 Field Effect Transistors for Terahertz Detection: Physics and First Imaging Applications *J. Infrared, Millimeter, and Terahertz Waves* **30** 1319–37.
[3] Bauer M, Venckevičius R, Kašalynas I, Boppel S, Mundt M, Minkevičius L, Lisauskas A, Valušis G, Krozer V and Roskos H G 2014 Antenna-coupled field-effect transistors for multi-spectral terahertz imaging up to 4.25 THz *Optics Express* **22** 19235.
[4] Mateos J and Gonzalez T 2012 Plasma Enhanced Terahertz Rectification and Noise in InGaAs HEMTs *IEEE Transactions on Terahertz Science and Technology* **2** 562–9.
[5] Rudin S, Rupper G, Gutin A and Shur M 2014 Theory and measurement of plasmonic terahertz detector response to large signals *J. Appl. Phys.* **115** 64503.
[6] Grasser T, Ting-Wei Tang, Kosina H and Selberherr S 2003 A review of hydrodynamic and energy-transport models for semiconductor device simulation *Proceedings of the IEEE* **91** 251–74.
[7] Lisauskas A, Pfeiffer U, Öjefors E, Bolìvar P H, Glaab D and Roskos H G 2009 Rational design of high-responsivity detectors of terahertz radiation based on distributed self-mixing in silicon field-effect transistors *J. Appl. Phys.* **105** 114511.
[8] Lisauskas A, Boppel S, Bauer M, Zdanevičius J, Matukas J, Krozer V and Roskos G 2014 Terahertz Detection with Field-effect Transistors: Intrinsic versus Device Sensitivity Limits *OSA Technical Digest (online)* Optical Sensors 2014 (Barcelona Spain: OSA) p SeTu5A.2.
[9] Lisauskas A, Bauer M, Ramer A, Ikamas K, Matukas J, Chevtchenko S, Heinrich W, Krozer V and Roskos H G 2015 Terahertz rectification by plasmons and hot carriers in gated 2D electron gases International Conference on Noise and Fluctuations (ICNF), (Xian: IEEE) pp 1–5.
[10] Gonzalez de la Cruz G and Gurevich Y G 2010 Carrier heating effects on transport phenomena in intrinsic semiconductor thin films *Revista mexicana de física* **56** 211–216.
[11] Bauer M, Andersson M, Zak A, Sakalas P, Čibiraitė D, Lisauskas A, Schröter M, Stake J and Roskos H G 2015 The potential for sensitivity enhancement by the thermoelectric effect in carbon-nanotube and graphene Tera-FETs *Journal of Physics: Conference Series* **647** 12004.

A Design of Terahertz Parallel Plate Dielectric Waveguide with Signal Line inserted for Ballistic Deflection Transistor Travelling Wave Amplifier

H. Wang[1, 2], R. Knepper[2], N. Hossain[1], P. Marthi[1], J-F. Milithaler[1] and M. Margala[1]

1. Department of Electrical and Computer Engineering, University of Massachusetts, 1 University Ave., Lowell, MA 01854
2. Department of Electrical and Computer Enigineering, Boston Univesity, 8 Saint Marys Street, Boston, MA 02215

Huan_Wang@student.uml.edu

Abstract. In this paper a new waveguide design is proposed to be implemented as part of Ballistic Deflection Transistor (BDT) Traveling Wave Amplifier Design. The BDT is designed to be operated in the Terahertz regime. Due to its relatively low transconductance (g_m=200μA/V), the entire structure will consist of ten stages, with 15 BDTs/stage, to reach a total gain of 30mA/V. In this case, the total length of the transmission line will be more than 400μm. We did the investigation for different structures and materials of the transmission line. For our Parallel Plate Dielectric Waveguide with Signal Line inserted (PPDWS) design, we are able to get an average loss of 0.46dB/mm at 0.8-1.4THz from ANSYS HFSS simulation. The return loss for input and output are better than -20dB at 0.8-1.7THz. Although it is designed for our future travelling wave amplifier, it can also be used for various other THz frequency applications.

1. Introduction

With the rapid growth in research and development of Terahertz (THz) devices, people are becoming more and more interested by applying THz devices into more applications such as imaging [1], telecommunications [2], biotechnology [3], astronomy [4]. Different kinds of devices have been introduced to be operated in THz regime for those purposes.

Our group has utilized a THz device called the Ballistic Deflection Transistor (BDT). This planar device is fabricated in an InGaAs/InAlAs heterostructure on InP substrate and consists of six terminals: a source, a top drain which pulls electrons, two output left and right drains and two lateral gates [5]. A negative bias applied on one gate will deplete the adjacent channel and, with the help of a strategically placed deflector, will steer electrons toward the opposite direction. This component provides a unique symmetric transfer characteristic which depends on geometry and polarization [6, 7]. More details can be found in [8].

Because of the low transconductance of the BDT, we would need at least ten stages, each stage comprised of 15 BDTs in parallel [9] to reach a gain of 3 mA/V per stage. A nearly lossless THz transmission line over 400 μm in length should be implemented into the amplifier design. However, for THz waveguide, neither conventional metal waveguides nor dielectric fibers are able to transmit this kind of very high frequency signal over a very long distance. The main issue is the high loss for the metal and high absorption efficiency for the dielectric materials in the THz regime [10]. The development of a THz transmission line will play a very important role in the designs of future THz applications.

In our own research, we have tried to implement the conventional coplanar waveguide (CPW) topology to design the THz waveguide, but the CPW is very lossy in the THz region and this is a key issue for THz CPW design. Over the last ten years, varieties of THz waveguide have been proposed such as non-radiative dielectric (NRD) waveguide [11], parallel plate waveguide (PPW) [12, 13], and parallel plate dielectric waveguide (PPDW) [14]. Both NRD and PPW are operated at higher transmission modes, but PPDW works at the fundamental TE_{10} mode. What's more, PPDW has a simpler structure and very low loss over a long distance in the THz region. However, the PPDW is not able to be directly connected with semiconductor devices. Thus, a new structure called Parallel Plate Dielectric Waveguide with an inserted signal conductor wire (PPDWS) is introduced.

2. Parallel Plate Dielectric Plate Design

The PPDWS is presented in Fig 1. The structure of the PPDWS consists of a rectangular dielectric strip with a height of *h* and width of *w*. The center dielectric strip is sandwiched between two metal plates as the ground planes. A metal conductor wire is inserted in the center of the dielectric strip.

The metal gold, with a high conductivity ($\sigma=4.42\times10^{7}$ S/m), high melting point, resistant to oxidation compared to silver and widely implemented as the material for high frequency applications has been chosen for metal planes and the signal line. The dielectric constant of the center strip should be greater than the outside dielectric in order to keep the radiation energy inside the dielectric strip without any leakage from the dielectric strip. Furthermore, in the fabrication process for this type of structure we need supporting material to fill the gaps to support the upper metal plane, leading to a

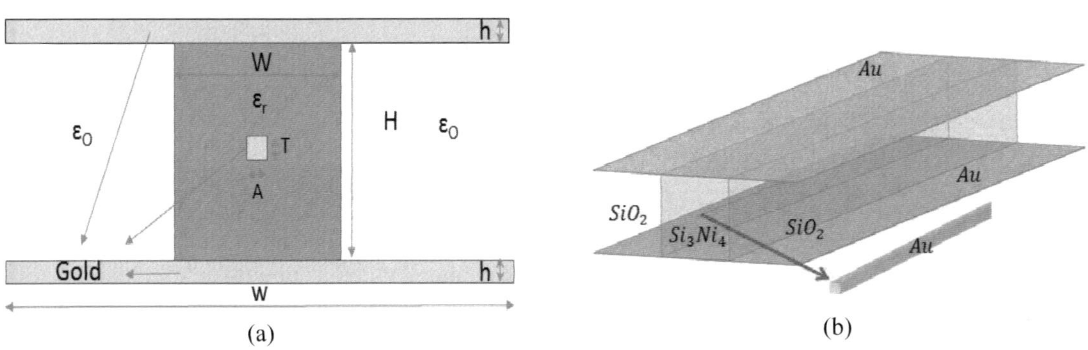

(a) (b)

Fig 1. (a) Cross section view of the PPDWS (b) 3D model of PPDWS

choice of Si_3N_4 (ε_r=9.5, tan$\delta=2\times10^{-5}$) for the center dielectric strip and SiO_2 (ε_o=3.7, tanδ=0.001) for the remainder.

The idea of the signal line derives from the conventional CPW design. In order to be connected with the BDT, we cannot use the PPDW directly. By taking the signal line from CPW and inserting it into the dielectric strip will directly connect the waveguide with the transistor's channel, which in this case is 300nm; as a result, the width of the signal line is equal to the channel width of the BDT (A=300nm). The thickness (T) of the signal line will be determined by the skin effect thickness given by the following equation:

$$\delta = \sqrt{\frac{2\rho}{2\pi f \mu_R \mu_0}} \tag{1}$$

f represents the operating frequency which is 1THz, ρ represents the resistivity of gold, μ_R is the relative permeability of gold and μ_0 is the permeability of free space equal to 4×10^{-7} H/m. Assuming $T \geq 3\delta$ which leads to $T \geq 297\ nm$, in this design 300 nm is applied as the thickness for the signal line. In Fig 1, the thickness of the dielectric (W=H=100 μm) will greatly influence the performance of the waveguide, is explained below in simulation results. The width of the metal plane is equal to 3W=300 μm, which will prevent the radiation leakage from two sides. The thickness of the metal plane is 500 nm, which is good for fabrication.

3. Simulation Results and Analysis

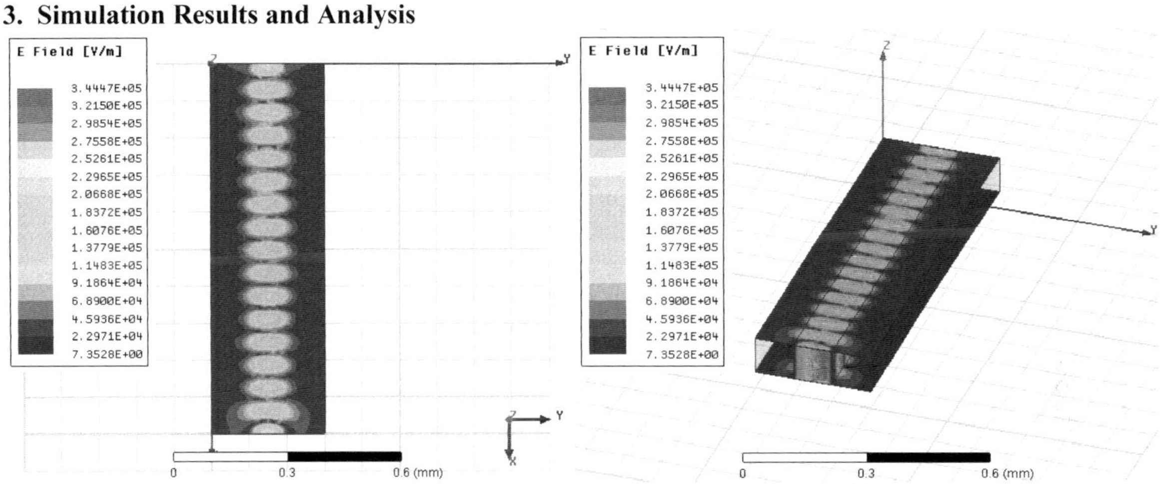

Fig 2. (a) Top view (b) Side view of Electric Field Distribution of PPDWS at 1THz (1mm long)

Fig 2 presents the field confinement of the PPDWS at 1THz. It clearly shows the TE10 mode in the electric field distribution. After propagating for the length of the transmission line, the waves inside almost remained unchanged, and the blue area shows that all the energy is locked inside the dielectric area and that no radiation is going outside.

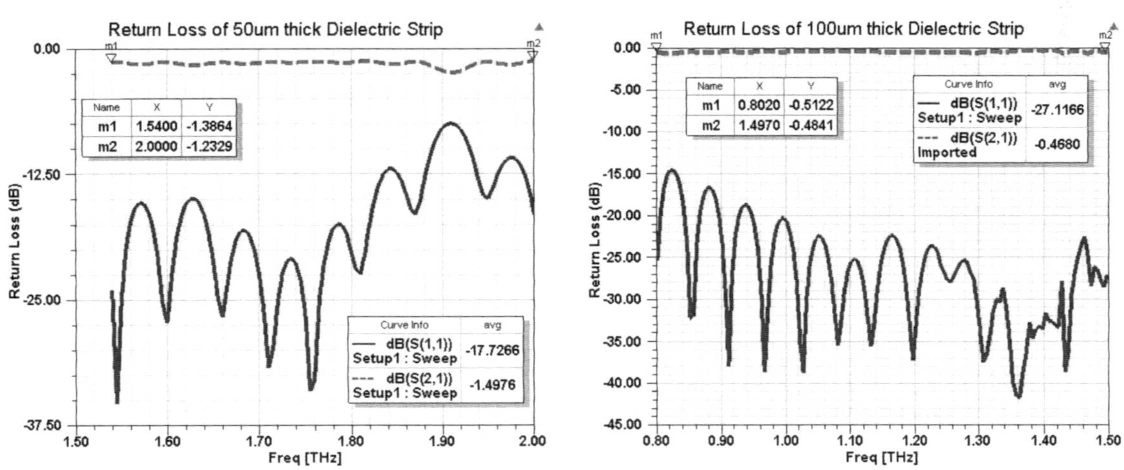

Fig 3. (a) Return Loss of a 50 μm thick dielectric strip at 1.54-2 THz (1 mm long) (b) Return Loss of the 100μm thick dielectric strip at 0.8-1.5 THz (1 mm long)

Fig 3. shows two plots of the S-parameter simulation results from HFSS. Comparing the results from (a) and (b), average total transmission loss for a 50μm thickness is 1.4976 dB/mm at 1.5-2 THz (best

IOP Conf. Series: Journal of Physics: Conf. Series **906** (2017) 012024 doi:10.1088/1742-6596/906/1/012024

bandwidth for the lowest transmission loss) with an average S11 of -17.7dB, while for a 100 μm thickness it is only 0.46 dB/mm at 0.8-1.5 THz, which is 1 dB better. We can state that the loss is greatly influenced by the thickness of the dielectric. By increasing the thickness of dielectric strip, we are able to reduce the loss of the transmission line, but it will also lead to the increase of the effective permittivity and power refinement [15]. When taken into consideration in the fabrication possibility, the 100 μm is the optimum thickness for the dielectric strip.

4. Conclusion

A new design of transmission line using the PPDWS technique is developed to be operated in the Terahertz band and integrated with THz BDT travelling wave amplifier. For a 100μm thick PPDWS has an average loss as low as 0.46 dB/mm at a broad frequency band at 0.8-1.5 THz. And the return loss at its input is below -20 dB for the frequency band. What's more, compared with the PPDW design, the PPDWS is able to be connected to the device directly. The operating frequency is also much higher and wider than with the PPDW structure. As a result PPDWS is an excellent candidate for the THz BDT travelling wave amplifier design and also has a great potential in other THz RF applications.

References

[1] J.F.Fedrici et al., "THz imaging and sensing for security applications, explosive weapons and drugs," *Semicond. Sci. & Tech.,* No.20, S266-280, 2005

[2] I. F. Akyildiz et al., "Terahertz band: Next Frontier for wireless commnuications," *Physical Commnuication* 12 (2014) 16-32

[3] X. Yang et al., "Biomedical Applications of Terahertz Spectroscopy and Imaging," *Trends in Biotech.*, Vol.34, No.10, Oct.2016

[4] C. Kulesa, "Terahertz Spectroscopy for Astronomy: From Comets to Cosmology," *IEEE Trans. Terahertz Sci. & Tech.*, Vol.1, No.1, September. 2011

[5] Q. Diduck et al., "A Room Temperature Ballistic Deflection Transistor For High Performance Applications," *International Journal of High Speed Electronics and Systems,* vol. 19, pp. 23-31, 2009.

[6] V. Kaushal et al., "A Study of Geometry Effects on the Performance of Ballistic Deflection Transistor," *IEEE Transactions on Nanotechnology*, vol. 9, pp. 723-733, 2010.

[7] M. Margala et al., "Current transport modeling and experimental study of THz room temperature ballistic deflection transistors," *J. Phys. Conf. Ser*, vol. 193, p. 4, 2009.

[8] J.-F. Millithaler et al. "Optimization of Ballistic Deflection Transistors by Monte Carlo Simulations," *J. Phys.: Conf. Ser.,* 647, 012066, (2015)

[9] M. Margal et al., "Ballistic deflection transistors and their application to THz amplification", *J. Phys.: Conf. Ser.* 647, 012020 (2015)

[10] H. Bao et al., "Dielectric tube waveguides with absorptive cladding for broadband, low-dispersion and low loss THz guiding", *Scientific Report* 5, Article No.7620 (2015)

[11] T. Yoneyma, "Nonradiative Dielectric Waveguide", *Infrered and Millimeter Waves*, Vol.11, Ch.2, K.J. ButtonEd. New York: Academic Press, 61-98 (1984)

[12] R. Mendis and D. Grischkowsky, "Undistorted guided-wave propagation of subpicosecond terahertz pulses", *Opt. Letter.* 26(11), 846-848(2011)

[13] R. Mendis, "Guided-wave THz time-domain spectroscopy of highly doped silicon using parrallel-plate-waveguides", Electron. Lett. 42(1), 19-21 (2006).

[14] L. Ye et al. "A novel broadband coaxial probe to parrlele palte dielectric waveguide transition at THz frquency", *Opt. Express.,* Vol.18, No.21 (2010)

[15] L. Cao et al. "Comparison and optimization of dispersion,and losses of planar waveguides on benzocyclobutene (BCB) at thz frequencies:coplanar waveguide (CPW), microstrip,stripline and slotline", *Progress in Electromagnetic Research B*, Vol.56, 161-183 (2013)

IOP Conf. Series: Journal of Physics: Conf. Series **906** (2017) 012025 doi:10.1088/1742-6596/906/1/012025

Control of quantum dot laser emission by coherent phonon wave packets

D. Wigger[1], T. Czerniuk[2], D. E. Reiter[1], M. Bayer[2], and T. Kuhn[1]

[1] Institut für Festkörpertheorie, Universität Münster, 48149 Münster, Germany

[2] Experimentelle Physik 2, Technische Universität Dortmund, 44221 Dortmund, Germany

E-mail: d.wigger@wwu.de

Abstract. Travelling coherent phonons can be actively used to manipulate the optical properties of semiconductor nanostructures on the picosecond time scale. Phonon wave packets that interact with a quantum dot (QD) ensemble can significantly vary the output intensity of a laser, which uses the QDs as active medium. Based on a recently developed theoretical model to describe this coupled phonon-QD-photon system, we here study how the laser response on phonon wave packets depends on several parameters, for example phonon pulse properties and laser pump rate.

1. Introduction and Theory

By using phonons to dynamically control the optical properties of nanostructures, one is pushing the field of opto-nano-electronics towards the growing fields of phononics and opto-mechanics. One example for this progress is the control of confined few-level systems by surface acoustic waves [1–3]. In contrast, we are using coherent phonon waves that travel through the bulk material and impinge vertically on an optically active layer. Several experiments have demonstrated the feasibility of the approach [4–7]. Here, we consider a single phonon pulse hitting a quantum dot (QD) layer, which serves as active laser medium. Figure 1(a) shows a schematic picture of the experimental setup. The basic laser system consists of a QD layer sandwiched between two distributed Bragg reflectors (DBR). The laser is optically pumped and emits through the front surface. The phonons are created by laser pulse heating a aluminum film at the back of the structure. The expansion of the metal within a few picoseconds launches coherent phonon wave packets towards the laser system.

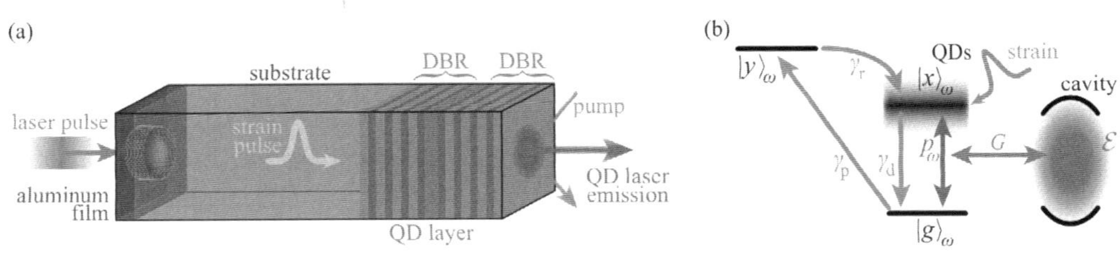

Figure 1. Schematic picture of the experiment (a) and the theoretical model (b).

We have recently developed a laser model, which includes the interaction of the QD excitons with the strain that corresponds to the displacement of the phonon field [8]. We have also shown that the model is capable of simulating the most important features of the laser emission dynamics [7]. By directly comparing measurements of the laser intensity with simulations within this model it was possible to distinguish between two mechanisms that govern the strain-QD interaction [7]: (i) The *adiabatic shift* of the ensemble, which changes the number of QDs in resonance with the laser cavity by shifting the QD ensemble in energy. (ii) The *shaking effect*, which dynamically brings highly occupied QDs into resonance with the cavity. While the former one can increase and decrease the output, the latter one always results in enhancements. A sketch of the employed three-level laser model including the strain interaction is depicted in Fig. 1(b). The full set of equations is derived in Ref. [8] based on the usual dipole, rotating wave, and slowly varying envelope approximations. Here, we model the spectral distribution of the QDs by a Gaussian of width Δ_{QD} centered around $\hbar\omega_0$:

$$n(\omega) = N_0 \exp\left\{-\left[(\hbar\omega - \hbar\omega_0)/\Delta_{\mathrm{QD}}\right]^2\right\} .$$

This is schematically depicted in Fig. 2(g), where also the initial detuning $\Delta_0 = \hbar\omega_{\mathrm{c}} - \hbar\omega_0$ with respect to the cavity frequency ω_{c} is introduced. The strain now acts on the QDs by shifting the exciton energies via $\hbar\omega_{\mathrm{x}}(t) = \hbar\omega_{\mathrm{i}} + D\eta(t)$, where $\hbar\omega_{\mathrm{i}}$ are the unstrained transition energies, D is the deformation potential coupling and $\eta(t)$ is the time dependent strain. For reasons of practicality we consider the relative pump rate $\Gamma_{\mathrm{p}} = \gamma_{\mathrm{p}} - \gamma_{\mathrm{d}}$ close to the laser threshold, given by the actual pump rate γ_{p} and the spontaneous decay rate γ_{d}.

2. Results

In the experiments solitonic phonon pulses develop, which have an approximately Gaussian time dependence of the strain [4]. Therefore we model the energy shift of the QD ensemble by

$$D\eta(t) = \eta_0^{\mathrm{D}} \exp\left[-(t/\tau_{\mathrm{s}})^2\right]$$

with a maximal energy shift of η_0^{D} and a duration of τ_{s}. In Fig. 2 we consider a pulse with $\eta_0^{\mathrm{D}} = 10$ meV and $\tau_{\mathrm{s}} = 5$ ps, a Gaussian QD ensemble of width $\Delta_{\mathrm{QD}} = 10$ meV and a pump rate of $\Gamma_{\mathrm{p}} = 10\ \mu\mathrm{s}^{-1}$. For each column in Fig. 2 a different initial detuning Δ_0 is chosen.

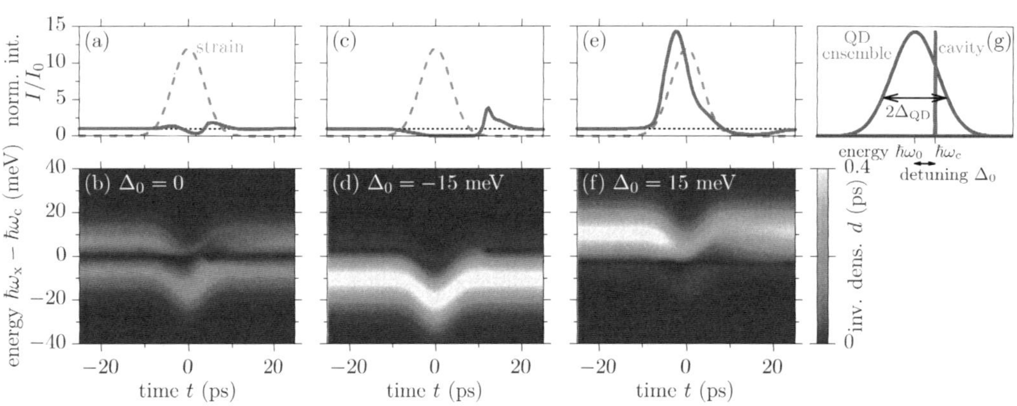

Figure 2. Top: Normalized intensity I/I_0 as a function of time. Strain pulse as dashed yellow line. Bottom: Inversion density d as a function of time and energy for a fixed cavity resonance. Different initial detunings for the columns. (g) Sketch of ensemble and cavity spectrum.

For $\Delta_0 = 0$ in Fig. 2(b) the inversion density $d = (x_\omega - g_\omega)n(\omega)$ of the ensemble ($g_\omega = \langle|g\rangle_\omega \langle g|_\omega\rangle$

and $x_\omega = \langle |x\rangle_\omega \langle x|_\omega \rangle$ are the occupations ground and exciton state, respectively) is shown as function of time t and energy $\hbar\omega_\mathrm{x} - \hbar\omega_\mathrm{c}$ with a fixed cavity resonance. The distribution exhibits the expected spectral hole in the center of the Gaussian ensemble. As a function of time the strain pulse leads to a shift of the ensemble, while the spectral hole stays at the energy of the cavity. The normalized laser intensity in Fig. 2(a) exhibits only minor variations, but we can see that, although the number of QDs in resonance with the cavity shrinks, the intensity slightly increases due to the shaking effect [8]. In the case of a negative detuning of $\Delta_0 = -15$ meV in Fig. 2(c,d) the initial detuning is increased by the strain pulse, which results in a full quench of the laser. After the strain pulse the laser switches on again, showing the characteristic relaxation oscillations. For the inverted case of a positive detuning in Fig. 2(e,f) the strain pulse leads to a reduction of the detuning, which brings more QDs into resonance with the cavity. This directly results in a strong enhancement of the laser intensity by a factor of almost 15. To sum up, for the positive detuning the strain pulse moves the ensemble towards the cavity mode resulting in enhancement, while for negative detuning the ensemble is shifted further away and the output is reduced.

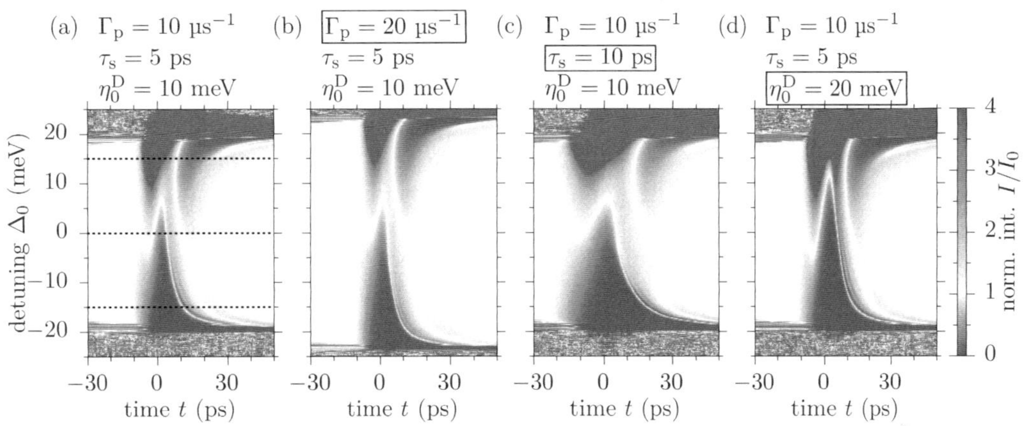

Figure 3. Normalized intensity as a function of time t and initial detuning Δ_0 for different parameters listed in (a-d). The dotted black horizontal lines in (a) mark the parameters from Fig. 2.

To extend the exemplary study of the initial detuning, in Fig. 3 the normalized intensity I/I_0 is plotted as a function of time t and initial detuning Δ_0. For Fig. 3(a) we take the parameters from Fig. 2. The basic features of enhancement (orange) and attenuation (blue) are restricted to detunings $|\Delta_0| \lesssim 20$ meV. For larger detuning values the pump rate is below the threshold and the output does not exceed the noise levels, yielding the sparkled pattern. We first want to focus on the blue quenching area at negative Δ_0. After this quenching part, the laser switches on again, with relaxation oscillations and therefore enhanced intensities (orange curved line). The orange areas for positive detunings represent enhancements, as the detuning is reduced by the strain. In these cases the dynamics start with the enhancement followed by an attenuation period, which arises from the relaxation to the equilibrium. Obviously, the orange area extends into the noise, which shows that the strain pulse can be used to initiate the laser switch-on. In the range of small detunings around $\Delta_0 = 0$ the two areas intersect and we find both, attenuations and enhancements, during the dynamics. This happens due to the interplay between adiabatic shift and shaking effect [8].
We now start from Fig. 3(a) and change one of the system parameters for (b-d) as written in the respective panel. For Fig. 3(b) the Γ_p is doubled to 20 µs^{-1}. This moves the noisy area to larger

detunings $|\Delta_0|$. Due to the stronger pumping also QDs with larger detuning are brought above the threshold and lead to lasing. Apart from this, no obvious changes appear. When we have a close look at the relaxation oscillations for negative detunings they become a little softer, which shows that the switch-on dynamics depends on the pump rate of the laser.
For Fig. 3(c) we increase the strain pulse duration to $\tau_s = 10$ ps. This basically stretches the time axis because the strain dynamics governs the laser output. One important but not very obvious feature is that the enhancement part around $|\Delta_0| = 0$ gets weaker. The reason is that this enhancement happens due to the shaking effect of the ensemble. This effect has a reduced influence, when the strain dynamics is slower [8], which is the case here.
For the example in Fig. 3(d) the amplitude of the strain shift is increased to $\eta_0^{\mathrm{D}} = 20$ meV. Comparing again this picture to panel (a) all features become stronger. On the one hand, the relaxation oscillations for negative detunings start at smaller values of Δ_0, which shows that the quench happens for smaller detunings. On the other hand, the attenuation area reaches far into the positive detuning regime because for these large strain shifts even starting with a positive Δ_0 results in a large negative detuning at the maximum of the strain pulse. Also the enhancement feature at $\Delta_0 = 0$ gets more pronounced. This can again be understood by the shaking effect. The effective strain shift gets faster when the duration stays the same but the amplitude gets increased.

3. Conclusions

In summary, we have presented a systematic study of several parameters of the coupled phonon-QD-photon system, namely the detuning between QDs and the cavity mode, the pump rate, the strain duration and the strain amplitude for the interaction with a Gaussian strain pulse influencing the laser output of a QD laser. All appearing features were explained by the two interaction mechanisms, namely the adiabatic shift and the shaking effect. This study is a first step towards an optimization of the system. Developing a detailed understanding of the influence of every tunable parameter paves the way to tailoring the actual experiment in a desired way.

Acknowledgments

The Dortmund team acknowledges the financial support by the Collaborative Research Center TRR 142. We are grateful for fruitful discussions with Andrey V. Akimov.

References

[1] Metcalfe M, Carr S M, Muller A, Solomon G S and Lawall J 2010 *Phys. Rev. Lett.* **105** 037401

[2] Gustafsson M V, Aref T, Kockum A F, Ekström M K, Johansson G and Delsing P 2014 *Science* **346** 207–211

[3] Schülein F J R, Zallo E, Atkinson P, Schmidt O G, Trotta R, Rastelli A, Wixforth A and Krenner H J 2015 *Nature Nanotech.* **10** 512

[4] Brüggemann C, Akimov A V, Scherbakov A V, Bombeck M, Schneider C, Höfling S, Forchel A, Yakovlev D R and Bayer M 2012 *Nature Photon.* **6** 30

[5] Czerniuk T, Brüggemann C, Tepper J, Brodbeck S, Schneider C, Kamp M, Höfling S, Glavin B A, Yakovlev D R, Akimov A V and Bayer M 2014 *Nature Comm.* **5** 4038

[6] Czerniuk T, Tepper J, Akimov A V, Unsleber S, Schneider C, Kamp M, Höfling S, Yakovlev D R and Bayer M 2015 *Appl. Phys. Lett.* **106** 041103

[7] Czerniuk T, Wigger D, Akimov A V, Schneider C, Kamp M, Höfling S, Yakovlev D R, Kuhn T, Reiter D E and Bayer M 2017 *Phys. Rev. Lett.* **118** 133901

[8] Wigger D, Czerniuk T, Reiter D E, Bayer M and Kuhn T 2017 *New J. Phys.* **19** 073001

IOP Conf. Series: Journal of Physics: Conf. Series **906** (2017) 012026 doi:10.1088/1742-6596/906/1/012026

Nanoscale engineering of photoelectron processes in quantum well and dot structures for sensing and energy conversion

X Zhang[1], V Mitin[1], A Sergeev[2], K Sablon[2], M Yakimov[3], S Oktyabrsky[3], J K Choi[4], G Strasser[5]

[1]State University of New York, Buffalo, New York, USA
[2]U.S. Army Research Laboratory, Adelphi, Maryland 20783, USA
[3]SUNY Polytechnic Institute, Albany, New York, USA
[4]Memory R&D Division, SK Hynix, Icheon-si, Gyeonggi-do, 467-701, Korea
[5]Center for Micro- and Nanostructures, TU Vienna, Vienna 1040, Austria

E-mail: xzhang49@buffalo.edu

Abstract. Advanced selective doping provides effective tool for nanoscale engineering of potential barriers and photoelectron processes in quantum well (QW) and quantum dot (QD) optoelectronic nanomaterials for IR sensing and wide band photovoltaic conversion. Photoelectron kinetics and device characteristics are investigated theoretically and experimentally. Asymmetrical doping of QWs is employed in a double QW structure for tuning electron transitions in QWs by voltage bias. These QW devices demonstrate bias-tunable multi-color detection and capability of remote temperature sensing. The QD structures with bipolar doping are proposed to independently control photocarrier lifetime (photocurrent) and dark current. The bipolar doping allows us to increase the height of nanoscale potential barriers around QDs without changing the electron population in QDs, which determines dark current. The QD devices with bipolar doping demonstrate significant enhancement of photocurrent, while dark current is close to that in corresponding reference devices with unipolar doping.

1. Introduction

With the advancement of material growth technology, semiconductor nanomaterials offer numerous opportunities for developing advanced optoelectronic devices. Nanostructures with quantum wells (QWs) and quantum dots (QDs) provide high degree of flexibility for engineering of band structure and nanoscale potential barriers. Selective doping in QW and QD nanomaterials allows us to control the nanoscale potential profile, created by charged QWs and QDs, and, in this way, to optimize photoelectron processes for enhanced sensing and photovoltaic conversion [1-4].

Here we present our recent results on advanced QW and QD nanomaterials with specific selective doping. In Section 2, we investigate asymmetrical doping of GaAs/AlGaAs double quantum wells (DQW) structures. Such doping enhances bias-induced charge redistribution between wells, which, in turn, modifies electron transitions in QWs. As a result, these DQW structures demonstrate significant tunability of spectral photoresponse characteristics, which can be effectively controlled by changing the polarity and/or magnitude of voltage bias. In Section 3, we report results of our studies of selective bipolar doping in QD structures. The bipolar doping combines the p-type doping in QD layers with the acceptors, N_A, and the n-type doping of inter-dot layers with donors, N_D. The number of free electrons in populated QDs is proportional to $N_D - N_A$. Population of QDs determines the dark current and, therefore, the dark current is also proportional to $N_D - N_A$. The potential barriers around dots are determined by the electron charge in QDs that is proportional to N_D. The potential barriers strongly suppress photoelectron capture and increase photoelectron lifetime.

2. Asymmetric doping of double quantum wells

Asymmetrically doped DQW structures were grown on semi-insulating GaAs wafer to permit backside illumination. The growth sequence started with a 500 nm undoped GaAs buffer layer, and was followed by a 800 nm heavily-doped ($N_D = 2\times10^{18}$ cm^{-3}) GaAs contact layer, 25 stages of the detection unit, a 400 nm heavily-doped ($N_D = 2\times10^{18}$ cm^{-3}) GaAs contact layer, and finally 5 nm InGaAs layer doped to 10^{19} cm^{-3}. One DQW unit is composed of a 6.5 nm GaAs layer doped by Si with sheet density 5×10^{11} cm^{-2} and 6.5 nm layer of undoped GaAs. They are separated by a 3.1 nm thick $Al_{0.2}Ga_{0.8}As$ barrier. The DQW units are separated by 50 nm $Al_{0.2}Ga_{0.8}As$ barriers. Grown structures were processed by standard photolithography, wet chemical etching, and metallization techniques.

2.1. Tunable characteristic of multi-color detection

In single quantum well infrared photodetector (QWIP) device, it is known that dark current at positive bias is usually higher than that at corresponding negative bias due to the dopant migration in the growth direction. In Fig. 1, dark current in our DQW devices is practically symmetrical with respect to applied bias. This is because the dopant migration is essentially compensated by asymmetrical doping in our DQW structures.

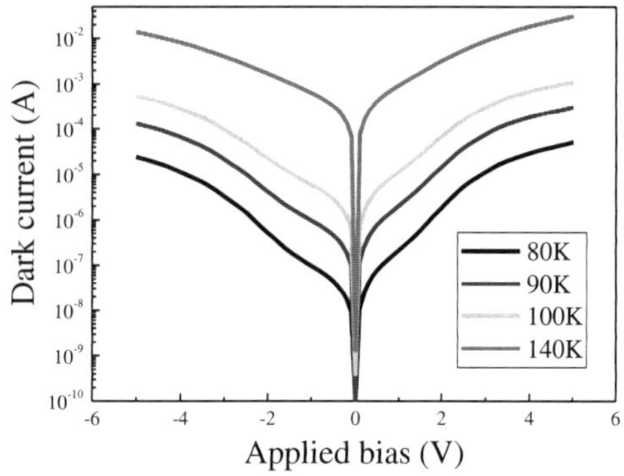

Figure 1. Dark current of QWIP based on asymmetrically doped DQW structures at 80K, 90K, 100K and 140K. They are measured by Keithley 2602 source-meter with grounded bottom contact.

The asymmetrical doping in DQW structure enhances bias-induced charge redistribution. As a result, these structures demonstrate significant tunability of electron transitions in DQW and corresponding tunability of spectral photoresponse. In Fig. 2, the detection peak varies from 7.5 μm to 11.1 μm when moderate voltage bias applies. It demonstrates that spectral photoresponse strongly depends on magnitude and polarity of applied bias due to induced electron charge redistribution. Tunable spectral characteristics are very attractive for adaptive IR sensing in complex environment.

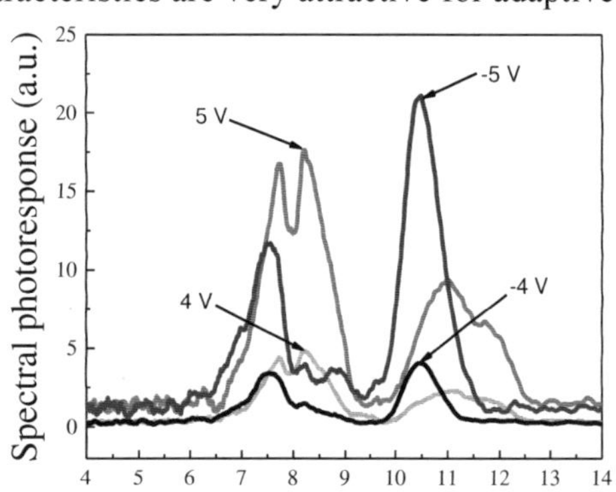

Figure 2. Spectral dependency of photoresponse in QWIP at T = 80K under positive bias 4 V, 5 V and negative bias -4 V, -5 V. Spectral photoresponse is recorded by VERTEX 70 FTIR spectrometer integrating SR570 low-noise current preamplifier.

2.2. Remote temperature sensing

In order to evaluate the capability of temperature sensing in our devices, we investigate the dependency of photocurrent on object temperature theoretically and experimentally. The modeling photocurrent is the integral of product of incident power and device responsivity over corresponding wavelength range. The experimental photocurrent is straight measured by current amplifiers using standard blackbody radiation. As shown in Fig. 3 (a), it demonstrates a good agreement between normalized modeling and experimental photocurrent, and they both strongly depend on object temperature.

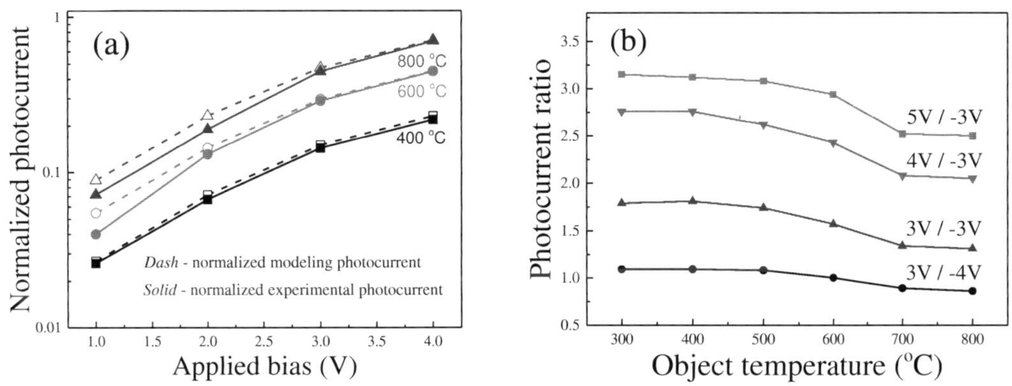

Figure 3. (a) Normalized modeling and experimental photocurrent at 400°C, 600°C and 800°C. (b) Photocurrent ratio as a function of object temperature.

Object temperature is not the only factor that determines thermal radiation. Emissivity and geometrical factor also affect its characteristic. Single-color photodetector cannot determine object temperature by simply measuring photocurrent, because it is difficult to eliminate the effects of emissivity and geometrical factor. However, multi-color photodetector is capable of determining object temperature because the ratio of photocurrents at two biases from the same device is only a function of object temperature, according to Eq. (1). It is also experimentally verified in Fig. 3 (b).

$$r(T)=\frac{I_{pc1}}{I_{pc2}}=\frac{\varepsilon g\int B(\lambda,T)\cdot R(\lambda,V_1)d\lambda}{\varepsilon g\int B(\lambda,T)\cdot R(\lambda,V_2)d\lambda}=\frac{\int B(\lambda,T)\cdot R(\lambda,V_1)d\lambda}{\int B(\lambda,T)\cdot R(\lambda,V_2)d\lambda} \quad (1)$$

3. Enhanced nanoscale potential profile around quantum dots

The key challenge in improving the performance of quantum dot IR photodetector (QDIP) is the trade-off between photocurrent and dark current. As two currents share the common path of electron transitions, high doping level increases the photocurrent, but inevitably leads to higher dark current. In order to independently control the photocurrent and dark current, we propose bipolar doping in QD structures. It combines p-type doping in QD layers with the acceptors, N_A, and n-type doping in inter-dot layers with donors, N_D.

The dark current is given by

$$J_{dc}=eN_{th}/\tau_{tr}=en_{th}S/\tau_{tr}\propto(N_D-N_A) \quad (2)$$

where n_{th} is thermally generated carrier, S is the device square and τ_{tr} is the electron transit time.

The photoelectron lifetime is exponentially increased due to potential barriers around negatively charged quantum dots

$$\frac{1}{\tau_l}=\pi n_D a^3\frac{1}{\tau_\varepsilon}\exp[-e\phi(Q)/kT]\propto\exp(-\frac{\phi(N_D)}{kT}) \quad (3)$$

where n_D is dot concentration, a is QD radius, $1/\tau_\varepsilon$ is inelastic electron-phonon relaxation rate, and ϕ is the potential at the dot boundary.

In the experiment, we design sample 4e0p ($N_D = 4$) with unipolar n-type doping outside of dots that supply on average 4 electrons per dot, and sample 7e3p ($N_D = 7$, $N_A = 3$) with n-type doping outside of

IOP Conf. Series: Journal of Physics: Conf. Series **906** (2017) 012026 doi:10.1088/1742-6596/906/1/012026

dots that supply on average 7 electrons pet dot and p-type doping in QD layers that bind about 3 electrons per dot to retain the same number of free electrons per dot. Results are presented in Fig. 4. In Fig. 4 (a), photocurrent in 7e3p is improved by 25% compared with 4e0p, but they have the close dark current. It is in good agreement with Eq. (2) and Eq. (3) where dark current is proportional to N_D - N_A and photoelectron lifetime (photocurrent) is proportional to N_D. In Fig. 4 (b), the voltage dependency of responsivity demonstrates the strong enhancement of photocurrent due to bipolar doping.

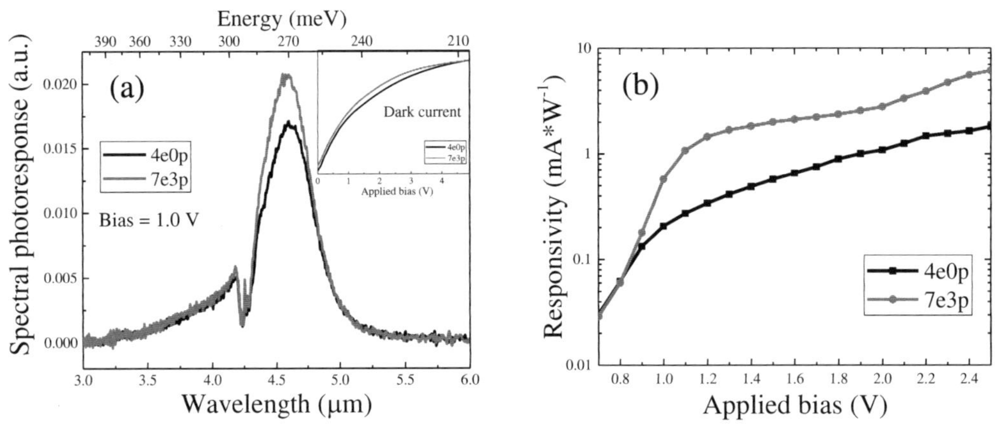

Figure 4. (a) Spectral photoresponse of 4e0p and 7e3p at 1.0 V and 80 K. The inset shows dark current. (b) Responsivity as a function of applied bias at 80 K.

4. Conclusions

We investigated the effects of nanoscale engineering for managing photoelectron processes in quantum well and quantum dot nanomaterials by employing selective doping. In our quantum well infrared photodetector devices, the asymmetrical doping between double QWs creates bias-induced charge redistribution between QWs. Due to the charge redistribution the bias-dependent photoresponse peak can vary from 7.5 μm to 11.1 μm depending on the polarity and/or magnitude of bias. Also, the photocurrent ratio is only a function of object temperature, which provides capability of remote temperature sensing. In quantum dot infrared photodetector (QDIP) structures, the bipolar doping is theoretically and experimentally proved to provide independent managing of photocarrier lifetime (i.e. photocurrent) and dark current. In particular, the device with 7e3p bipolar doping shows strongly enhanced photocurrent and practically the same dark current as the device with 4e0p unipolar doping. Thus, due to longer photoelectron lifetime the bipolar doping increases the sensitivity of QDIPs.

References

[1] K A Sablon, J W Little, V Mitin, A Sergeev, N Vagidov and K Reinhardt, "Strong enhancement of solar cell efficiency due to quantum dots with built-in charge", *Nano Letters* 11 (6), 2311-2317 (2011).

[2] A Varghese, M Yakimov, V Tokranov, V Mitin, K Sablon, A Sergeev and S Oktyabrsky, "Complete voltage recovery in quantum dots solar cells due to suppression of electron capture", *Nanoscale* 8 (13), 7248-7256 (2016).

[3] X Zhang, V Mitin, G Thomain, T Yore, Y Li, J K Choi, K Sablon and A Sergeev, "Reconfigurable infrared photodetector based on asymmetrically doped double quantum wells for multicolor and remote temperature sensing", *Future Trends in Microelectronics: Journey into the Unknown*, 327-336 (2016).

[4] K Sablon, A Sergeev, X Zhang, V Mitin, M Yakimov, V Tokranov and S Oktyabrsky, " Selective doping of quantum dots nanomaterials for managing intersubband absorption, dark current, and photoelectron lifetime", *MRS Advances* 2 (14), 759-766 (2017).

IOP Conf. Series: Journal of Physics: Conf. Series **906** (2017) 012027 doi:10.1088/1742-6596/906/1/012027

Dynamics of copropagating edge states in a multichannel Mach-Zender interferometer

L. Bellentani[1], **A. Beggi**[1], **P. Bordone**[1,2], **A. Bertoni**[2]
[1]Dipartimento di Scienze Fisiche, Informatiche, Matematiche
Universitá degli studi di Modena e Reggio Emilia, via Campi 213/A 41125 Modena, Italy
[2]CNR-Istituto Nanoscienze, Modena, Italy

E-mail: laura.bellentani@unimore.it

Abstract. We study numerically a multichannel electronic Mach-Zender interferometer, where an orthogonal magnetic field produces edge states. Our time-dependent model is based on the split-step Fourier method and describes the charge carrier as a Gaussian wavepacket of edge states, whose path is defined by split-gate induced potential profiles on the 2DEG at filling factor $\nu = 2$. We analyse a beam splitter with $\sim 50\%$ inter-channel mixing and obtain Aharonov-Bohm oscillations in the transmission probability of the second channel.

1. Introduction

Topologically protected edge states (ES) in the quantum Hall regime are able, in principle, to encode and process quantum information via electron quantum optics devices [1, 2, 3]. The realization of such semiconductor quantum logic gates requires suitable electrostatic potential profiles for the dynamical control of flying qubits, which is the target of our work.
In detail, we model a multichannel Mach-Zender interferometer (MZI) with bulk filling factor $\nu = 2$ [1, 4], where two parallel spin-degenerate edge states are forced to follow different paths. In contrast to the Corbino interferometer [3], we adopt the geometry proposed in Ref.[4], which allows concatenation in series and a smaller loop area, therefore reducing the effect of phase averaging. Instead of simulating delocalized plane-waves, we study the single-particle dynamical properties of the MZI describing the travelling charge as a strongly localized wavepacket (WP) of edge states [5, 6]. In analogy to the single channel MZI [7], our results focus on the measurement of Aharonov-Bohm (AB) oscillations, comparing numerical data to a simplified analytical model.

2. Physical system

Our model describes a 2DEG on the x-y plane, where the integer quantum Hall effect is generated by a uniform and orthogonal magnetic field B. A local confining potential $V(x)$ - constituting the edges of the MZI - is introduced to form edge states, chiral and quasi-adiabatic channels following the boundary between low-potential and high-potential regions of the device. With these assumptions, neglecting electron-electron interaction, the wavefunction of the single-particle Hamiltonian can be rewritten as $\Psi_n(x, y, k) = \phi_{n,k}(x)e^{iky}$. The exponential factor e^{iky} defines a travelling state along the y direction with momentum k, while $\phi_{n,k}(x)$ is the eigenstate of the 1D effective Hamiltonian with energy E_n at fixed k.

IOP Conf. Series: Journal of Physics: Conf. Series **906** (2017) 012027 doi:10.1088/1742-6596/906/1/012027

The study of dynamical properties in our time-dependent approach requires the description of the electronic state with a localized wavefunction. Therefore, the ESs are combined with Gaussian weights $F_\sigma(k) = \sqrt{\sigma}(2\pi)^{\frac{-3}{4}}e^{-\sigma^2(k-k_0)^2}e^{-iky_0}$ in the function $\Psi_n^o(x,y) = \int dkF_\sigma(k)\phi_{n,k}(x)e^{iky}$, which represents a minimum-uncertainty initial state localized at $(x_0 = -\hbar k_0/eB, y_0)$ with group velocity $v_g = \hbar k_0/m^*$.

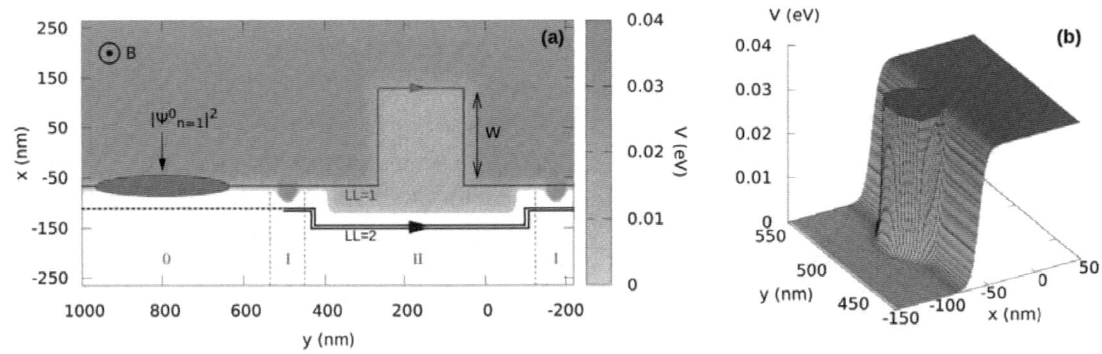

Figure 1. The two-channel MZI device. **a)**Top view of the device with the intended path of the two channels (single line for n=1 and double line for n=2) and initial electronic density probability ($|\Psi^0_{n=1}|^2$). Region *0*: injection; Region *I*: beam splitters; Region *II*: step potential. **b)** Potential profile adopted for the beam splitter.

After the injection region, where the potential varies only along x, V presents also a dependence on y [4]. Specifically, the paths of the ESs, together with two BSs, form a MZI. In order to mimic the smooth potential induced by realistic split gates, we adopt a Fermi-function shape for the transition between regions with different potential (i.e. at different filling factor). In contrast to our previous work [7], a bulk filling factor $\nu = 2$ characterizes the low-potential region, where the first two Landau levels ($n = 1, 2$) are available.
The complete device is plotted in Fig.1(a). Firstly, the wavepacket (initialized in $n = 1$) impinges on the beam splitter (BS) modeled with the sharp potential dip reported in Fig.1(b). Here, the scattering process coherently mixes the wavefunction in the two copropagating channels with different index n. Then, a suitable potential profile forces a different path for the two channels and the accumulation of a relative phase. Finally, the two parts of the wavefunction are recollected on a second beam splitter, where they interfere. At the end of the MZI the selective measurement of the transmission probability is performed through an additional mesa and an imaginary potential, which adsorbs only the channel with $n = 1$. Consequently, the norm of the final wavefunction coincides with the total transmission probability P_{21} from the first to the second channel, whose amplitude shows the characteristic oscillations induced by the interference.

3. Numerical simulations

In order to analyse the particle dynamics, we solve the time-dependent Schrödinger equation with the split-step Fourier method [7], using a time step $\delta t = 10^{-16}s$. The Gaussian WP of ESs is initialized in $n = 1$, $\sigma = 60nm$ and $k_0 = -0.387nm^{-1}$.

As a first step, we define a proper geometry of the potential $V(x, y)$ to obtain the BS. Following Ref.[4, 8, 9], such gate can be represented by a sharp potential profile, which splits elastically the incoming wave in the first two Landau levels. Indeed, the scattering redistributes the k components of the wavepacket in the available states, with a probability measured by the transmission coefficient $t^{BS}_{f,i}$ (i=1,2 for the initial state and f=1,2 for the final one). Such a parameter can be controlled by tuning the energy composition of the wavepacket, the magnetic field intensity B and the shape of the potential barrier. Regarding the sharpness of the local potential, numerical calculations [8] show indeed that spatial inhomogeneities on a scale smaller than the magnetic length are required to split efficiently the impinging state. After testing various potential profiles, we achieve a $t^{BS}_{2,1}$ coefficient of $(50 \pm 1)\%$ with the shape plotted in Fig.1(b), at B=5T. We simulate the scattering of the initial WP in both the first and the second Landau level, with essentially symmetric results. Moreover, we extracted the energy dependence of the transmission coefficients t^{BS}_{fi}, observing a quasi-flat dispersion in our energy bias, which ensures a coherent 50% mixing of the two channels without the resonant condition [10].

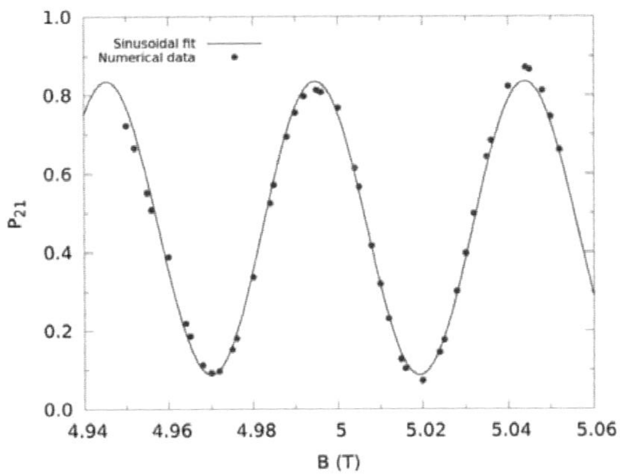

Figure 2. Transmission probability from $n = 1$ to $n = 2$ at the end of the device as a function of B simulated numerically (dots) and fit with eq.(1) (solid line).

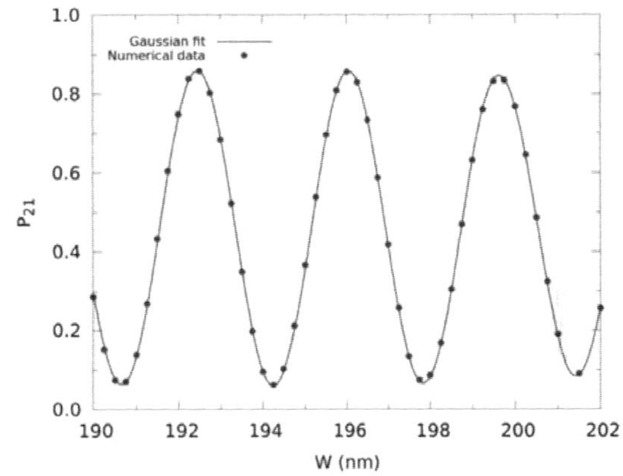

Figure 3. Transmission probability from $n = 1$ to $n = 2$ at the end of the device as a function of W simulated numerically (dots) and fit with eq.(2) (solid line).

Once the coherent superposition is realized, the mesa in Fig.1(a) splits the beam in two parts following different paths. Indeed, such potential step shifts upwards the band structure [4], forcing the two Landau levels to be filled at different wavevectors. Considering the relation $x = -\frac{\hbar k}{eB}$, this implies a different localization of the two channels in the x direction. Therefore, the two paths can be tuned by reshaping the local potential, and a length mismatch introduces a phase difference. To this aim, we design a step potential with smooth edges and avoid an undesired additional inter-channel mixing. Furthermore, our time-dependent approach entails a particular attention to the shape of the mesa. Indeed, this must ensure the simultaneous recollection of the two localized wavepackets at the second beam splitter, taking into account the different group velocities.

Table 1. Parameters for the transmission probability $P_{21}(W)$: a) from a fit of numerical simulation with eq.(2) and b) from the analytical model.

	A	A_0	ν_{21}	k_x (nm^{-1})	Σ (nm)
a)Fit of simulations	0.400±0.004	0.460±0.002	0.87	1.750	21.7±0.5
b)Analytical model	0.5	0.5	1	1.9	18.3

IOP Conf. Series: Journal of Physics: Conf. Series **906** (2017) 012027 doi:10.1088/1742-6596/906/1/012027

4. Results and discussion

We simulate the interference process varying both the magnetic field intensity and the length mismatch of the two paths at the mesa. We obtain AB oscillation of the transmission probability P_{21}, plotted in Fig.2 and Fig.3 for a variation of B and W (width of the mesa in Fig.1(a)) respectively. The numerical results are fit by the following functions:

$$P_{21}(B) = A_0 + Ae^{(-\frac{(W-W_0)^2}{2\Sigma^2})} cos(k_B B + \Phi_1), \quad (1)$$

$$P_{21}(W) = A_0 + Ae^{(-\frac{(W-W_0)^2}{2\Sigma^2})} \cos(k_x W + \Phi_0), \quad (2)$$

which describe the total transmission probability in our MZI for the initial Gaussian WP of ESs previously described. The results of the Gaussian fit in eq.(2) are finally compared to the predictions of our simplified theoretical model [7] in Tab.1.

Concerning the magnetically-driven interference, we observe that the oscillations do not form a perfect sinusoid, but the amplitude slightly increases with the magnetic field intensity. We relate this feature to the actual magnetic dependence of the scattering at the BS and at the mesa. Indeed, we must stress that eq.(1)-(2) are calculated assuming $t^{BS}_{1,2} = 0.5$, without including the dependence on energy and B. On the other hand, in Fig.3 the variation of the oscillation amplitude is induced by the Gaussian shape of the wavefunction, which is a unique feature of the time-dependent approach. In detail, when the length mismatch between the two paths approaches σ, the overlap between the two wavefunctions is reduced and the oscillation amplitude becomes weaker [11]. Moreover, in contrast to theoretical predictions, we achieve a maximum visibility $\nu_2^{max} \approx 87\%$ in place of unity. Different factors contribute to this reduction. Firstly, our theoretical model neglects the energy dependence of the dynamic phase and transmission amplitudes, assuming the impinging wavefunction equally scattered in the two available states. Additionally, despite its smoothness, the step potential induces a small inter-channel mixing (around 2%), which could be eventually reduced by shifting upwards the height of the mesa or increasing the smoothness of the barrier. Finally, we neglect the partial filling of the third Landau level, which is an available state in the selected energy bias.

In conclusion, we showed that the oscillating transmission of a two channel MZI with realistic dimensions can be obtained by solving numerically the time-dependent Schrödinger equation with the corresponding potential landscape. We showed how specific features of the transmission characteristic are linked to the localization of the propagating carrier. Our single-particle simulations validate this system for the implementation of single-qubit logic gates, and, in perspective, the simulation of two interacting carriers should expose its ability to create two-qubits gates.

References

[1] Neder I, Heiblum M, Mahalu D, and Umansky V 2007 *Phys. Rev. Lett.* **98** 036803
[2] Roddaro S, Chirolli L, Taddei F, Polini M and Giovannetti V 2013 *Phys. Rev.* B **87** 075321
[3] Deviatov E V 2013 *AIP Low. Temp. Phys* **39** 7
[4] Giovannetti V, Taddei F, Frustaglia D, and Fazio R 2008 *Phys. Rev.* B **77** 155320
[5] Gaury B, Weston J and Waintal X 2014 *Phys. Rev.* B **90** 161305
[6] Kataoka M, Johnson N, Emary C, See P, Griffiths J P, Jones G A C, Farrer I, Ritchie D A, Pepper M and Janssen T J B M 2016 *Phys. Rev. Lett* **116** 126803
[7] Beggi A, Bordone P, Buscemi F and Bertoni A 2015 *J. Phys.: Cond. Matt.* **27** 475301
[8] Palacios J J and Tejedor C 1991 *Phys. Rev.* B **45** 9059
[9] Paradiso N, Heun S, Roddaro S, Venturelli D, Taddei F, Giovannetti V, Fazio R, Biasiol G, Sorba L, and Beltram F 2011 *Phys. Rev.* B **83** 155305
[10] Karmakar B, Venturelli D, Chirolli L, Giovannetti V, Fazio R, Roddaro S, Pfeiffer L N, West K W, Taddei F, and Pellegrin V 2015 *Phys. Rev* B **92** 195303
[11] Haack G, Moskalets M, Splettstoesser J and Buttiker M 2011 *Phys. Rev.* B **84** 081303

Institute of Physics
Dirac House, Temple Back
Bristol BS1 6BE UK

ISSN: 1742-6588
ISBN 978-1-5108-5037-8

CURRAN ASSOCIATES INC.
proceedings.com